Genomic Sequence Analysis for Exon Prediction Using Adaptive Signal Processing Algorithms

Genomic Sequence Analysis for Exon Prediction Using Adaptive Signal Processing Algorithms

Md. Zia Ur Rahman and Srinivasareddy Putluri

CRC Press
Taylor & Francis Group
Boca Raton London New York

CRC Press is an imprint of the
Taylor & Francis Group, an **informa** business

First edition published 2021
by CRC Press
6000 Broken Sound Parkway NW, Suite 300, Boca Raton, FL 33487-2742

and by CRC Press
2 Park Square, Milton Park, Abingdon, Oxon, OX14 4RN

Library of Congress Cataloging-in-Publication Data
Names: Rahman, Md. Zia Ur (Muhammad Zia Ur), author. | Putluri, Srinivasareddy, author.
Title: Genomic sequence analysis for exon prediction using adaptive signal
processing algorithms / Md. Zia Ur Rahman, Srinivasareddy Putluri.
Description: First edition. | Boca Raton : CRC Press, 2021. |
Includes bibliographical references and index. | Summary: "This book addresses
the issue of improving the accuracy in exon prediction in DNA sequences
using various adaptive techniques based on different performance
measures which is crucial in disease diagnosis and therapy. Six chapters
are presented"— Provided by publisher.
Identifiers: LCCN 2020052479 (print) | LCCN 2020052480 (ebook) |
ISBN 9780367615802 (hardback) | ISBN 9780367618551 (ebook)
Subjects: LCSH: Genomics. | Nucleotide sequence—Mathematical models. |
Exons (Genetics) | Adaptive signal processing—Mathematics. |
Adaptive filters—Mathematical models.
Classification: LCC QH447 .R32 2021 (print) | LCC QH447 (ebook) |
DDC 572.8/6—dc23
LC record available at https://lccn.loc.gov/2020052479
LC ebook record available at https://lccn.loc.gov/2020052480

ISBN: 978-0-367-61580-2 (hbk)
ISBN: 978-0-367-61857-5 (pbk)
ISBN: 978-0-367-61855-1 (ebk)

Typeset in Times
by codeMantra

Contents

Authors

Dr. Md. Zia Ur Rahman received his M.Tech. and Ph.D. degrees from Andhra University, Visakhapatnam, India. He is currently a professor in the Department of Electronics and Communication Engineering, Koneru Lakshmaiah Educational Foundation (K. L. University), Guntur, India. His current research interests include adaptive signal processing, biomedical signal processing, array signal processing, MEMS, and nanophotonics. He has published more than 100 research papers in various journals and proceedings.

He is currently serving on various editorial boards as the Editor-in-Chief, an Associate Editor, and a reviewer of publishers like the IEEE, Elsevier, Springer, IGI, American Scientific Publishers, and Hindawai.

Dr. Srinivasareddy Putluri is currently a software engineer at Tata Consultancy Services Ltd., Hyderabad. He received his Ph.D. degree (Genomic Signal Processing using Adaptive Signal Processing algorithms) from the Department of Electronics and Communication Engineering, Koneru Lakshmaiah Educational Foundation, (K. L. University), Guntur, India. His research interests include genomic signal processing and adaptive signal processing. He has published 15 research papers in various journals and proceedings. He is currently a reviewer of publishers like the IEEE Access and IGI.

1 Introduction

1.1 GENOMICS ENGINEERING

In the past few decades, engineering in genomics has become a vigorous area of study that has attracted massive attention from the digital signal processing (DSP) research community. Some of its primary goals are analysis of deoxyribonucleic acid (DNA) sequences, protein structure modeling, locating exon segments in gene sequences, and finding correlations between various gene expression profiles. Genomic sequence analysis can be classified into two well-researched areas: DNA sequence analysis and microarray analysis. Signal processing in genomics is a novel field of research that combines DSP methods for a superior gene data analysis [1]. This book discourses a substantial challenging issue, namely, exon prediction in the DNA sequences of *Homo sapiens* in the area of genome sequence analysis. With the recent completion of numerous very large-scale genomic sequencing projects (e.g. human, *Caenorhabditis elegans*) and the need for a future scope of learning about different novel gene sequences, this issue is considered on major real significance. This issue arises primarily as the nature of genes in a genome is not continuous. Eukaryotic genes are further categorized into coding regions termed as exons and also other sections named as introns. Moreover, both these sections comprise the major part of a genome. The exon sections occupy only 3% of the human DNA.

The narrow accuracy of the various available data-driven software programs for gene prediction such as AUGUSTUS [2], FGENESH [3], GeneID [4], GeneMark.hmm [5], Genie [6], GENSCAN [7], HMM-gene [8], MORGAN [9], and MZEF [10] is an indication for improvement in the accuracy of gene prediction. The effectiveness of different gene finding programs is presented in Ref. [11]; also the use of Bayesian gene prediction, as illustrated in Ref. [12], has been put forward to use non-traditional methods, and thus this area is worth to investigate further. This can be dealt thru inspecting the roles of specific DSP procedures and developing different novel adaptive techniques towards improving the effectiveness of algorithms for exon identification. DSP-based techniques remain prominent as genomic data do not need any training, unlike prevailing data-driven methods.

1.2 DNA SEQUENCE STRUCTURE

The key info that is essential for building and also maintaining an organism's structure is present in a complex molecule known as DNA. It is comprised of genetic codes formed with four nucleotide bases: adenine (A), cytosine (C), guanine (G), and thymine (T). Phosphates along with sugar molecules are linked to every base, and all these are combined to form a nucleotide. The need for analyzing using DSP methods remains crucial due to diverse behaviors exhibited by DNA signals in different 'amplitudes' and 'times' (i.e. the base pair domain).

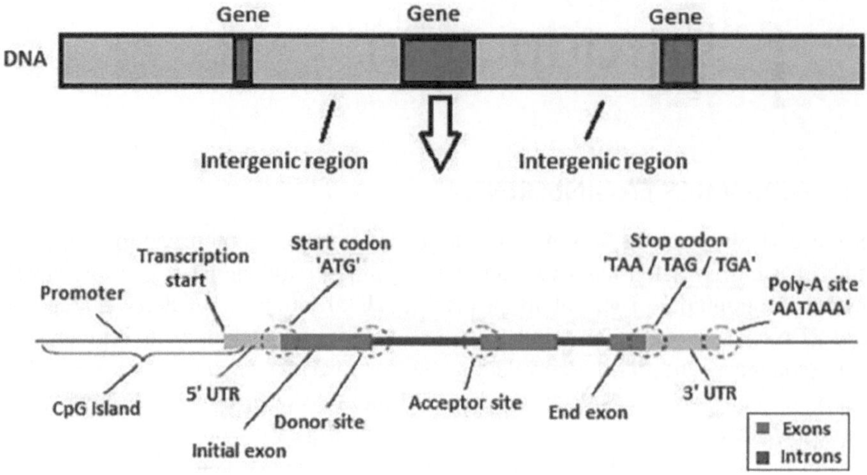

FIGURE 1.1 DNA with genic and intergenic regions.

Converting these alphabetic bases to distinct numeric sequences aids new beneficial applications of DSP to solve diverse issues connected to DNA like gene identification. DSP elucidates this task with an improved accuracy and also less complexity [13]. This work is highly cross-disciplinary in nature while basic subject matter is biological and final results obtained are of biological interest, techniques from other fields like DSP are greatly used.

The precise prediction of exon/intron sections, splice sites of acceptor/donor, additional segments, and also genetic signals of DNA with genic and intergenic sections is depicted in Figure 1.1. These positions relevant to 5′ and 3′ end points were accustomed in linking exon segments present on two sides of introns termed as splicing. A novel ailment-free computation method is presented in Ref. [14] for calculating the distance between a pair of sequences. In this work, the roles of DSP techniques in locating exon segments of genomic sequences are examined, and novel adaptive exon predictor (AEP)-based methods are developed for this purpose. Prevalent DSP techniques for exon prediction remain less accurate due to their lone dependency on period-3 identification, and they are also not good in terms of exon locating ability, computational complexities, and convergence behavior.

In this book, several methods of DNA symbolic-to-numeric representation for extraction of exon segments are reviewed and a Voss binary representation for conversion of alphabetic DNA to binary notation is also presented. The existing least mean squares (LMS) and new AEP-based techniques are then evaluated using real gene datasets taken from NCBI gene databank using various performance metrics computed at nucleotide level. We believe that the proposed novel DSP-based adaptive techniques are effective in gene identification, and also their ensuing AEPs offer improved gene prediction accuracy, better convergence performance, and considerably lower computational complexity than that offered by the existing LMS methods.

1.3 MOTIVATION FOR THE WORK

As observed thru different findings and an extensive survey of literature, exon prediction remains as a crucial step in disease diagnosis. Almost one in six deaths in the world is due to cancer, and thus, cancer has become the second most common cause of death. As per WHO statistics, cardiovascular and dreadful diseases like cancer are leading causes of death worldwide. Non-communicable diseases (NCDs), such as cardiovascular diseases (17.9 million), cancers (around 9 million), respiratory diseases (3.9 million), and diabetes (1.6 million), kill 41 million individuals every year worldwide.

As per World Cancer Report 2014, 8.2 million deaths are due to cancers. As per World Cancer factsheets from WHO by IARC in 2018, cancer deaths from both sexes of all ages increased to 9.5 million.

Every day in India, as many as 2,500 people die from tobacco-related illnesses. Tobacco is the most important risk factor for cancer and responsible for approximately 22% of cancer deaths. Proper diagnosis plays a crucial role for treating any disease. The main aim of this book is to study as well as create several effective AEPs based on DSP to identify exon areas accurately in the genomic sequences of *Homo sapiens* considered from the NCBI genome database.

1.4 OBJECTIVES

To study and also develop several capable AEPs using DSP techniques to accurately locate exon segments in the genomic sequences of *Homo sapiens* taken from the NCBI gene databank is the main objective of this book.

This wide objective includes all the below goals:

- To disparagingly analyze gene sequence data, biological literature, and also additional resources available online for gene identification and annotation of *Homo sapiens.*
- To study the role of specific DSP techniques in gene prediction problems in this area.
- To analyze genomic sequences and mapping the alphabetic DNA to binary sequence.
- To develop and review new DSP-based AEPs to identify exon locations in a genomic sequence.
- To investigate and rigorously evaluate, at nucleotide level, a wide range of the existing and new DSP methods to identify exon segments thru real gene datasets of the NCBI genome repository.
- To develop and analyze AEPs for prediction of exon locations in genomic sequences using signed LMS-based adaptive algorithms.
- To develop and analyze hybrid AEPs with normalization-based algorithms and its signed variants for better convergence, prediction, computational complexity, sensitivity, specificity, and precision.
- To extend the exon prediction phenomenon with hybrid versions of normalization-based logarithmic algorithms and its signed variants for better convergence and computational complexity.

1.5 MOLECULAR BASIS FOR GENOMIC INFORMATION

DNA remains to be the molecular source of genomic data that is present within the arrangement of its nucleotide bases as linear chains of subunits. The DNA characteristic that defines the molecule is the intertwining of the two strands as a right-hand double helix [15]. Every nucleotide is comprised of three portions: a sugar with five carbon atoms, a group of phosphates, and a base with nitrogen atoms. The resulting four probable bases of a nucleotide include adenine (A), guanine (G), cytosine (C), and thymine (T). However, thymine is substituted by uracil (U) in numerous viruses that have ribonucleic acid (RNA) as their building block. Distinct nucleotide bases within DNA remains associated by bonds between sugar phosphates, thus forming a lengthy chain with a single dimension by double dissimilar end points, i.e. 3′ end (downstream) and 5′ end (upstream). Consequently, this DNA series is denoted by a string of four alphabetic letters: A, G, C, and T.

1.5.1 Understanding the Genome

One or more constant sections of chromosomes are present in the complete genome of an organism. Elements for regulation, several genes, and additional prevailing sequences are contained in a chromosome.

A diverse amount of chromosomes may be present in distinct organisms. For instance, 46 chromosomes are present in each cell of *Homo sapiens*, while this amount is 40 in mice cells. Table 1.1 presents the total number of chromosomes and genome size for selected species as in Ref. [16]. All living organisms are categorized into eukaryotes and prokaryotes. The organisms that have no nucleus in their cells are prokaryotes, whereas eukaryotes are single- or multicell organisms whose cells contain a nucleus. Also, just one chromosome is generally part of prokaryotes, whereas two matched sets of chromosomes are part of eukaryotes, single set from each parent. Blue-green algae and bacteria are examples for prokaryotes, whereas humans are example for eukaryotes [17].

1.5.2 Building Blocks of DNA

Formerly, the F. C. H. Crick and J. D. Watson proposed that DNA is a double-helix model as presented in Figure 1.2 in Ref. [18], and ensuing fascinating characteristics for the analysis of genes in DNA structure are presented in Ref. [19]:

i. A double helix at right formed by two DNA strands twisted around a center axis.
ii. Both components remain antiparallel in definite reverse directions.
iii. Hydrogen bonds form nitrogen bases part of opposite strands in DNA.
iv. Pyrimidine in one chain is linked with purine from another chain; i.e. nucleotide base A is associated to T, while G is associated to C, and vice versa.
v. Pairs of pyrimidine and purine remain counterpart to each other, thereby resulting in dual chains of one molecule of DNA. Consequently, if the sequence 5' – CATTGCCAGT – 3' arises from a single chain, the sequence of the other strand should be 5' – ACTGGCAATG – 3', i.e.:

TABLE 1.1
Genome Sizes of Certain Species

Species	Number of Chromosomes (diploid)	Genome Size (base pairs)
Bacteriophage (virus)	1	5×10^4
Escherichia coli (bacterium)	1	5×10^6
Saccharomyces cerevisiae (yeast)	32	1×10^7
Caenorhabditis elegans (worm)	12	1×10^8
Drosophila melanogaster (fruit fly)	8	2×10^8
Homo sapiens (human)	46	3×10^9

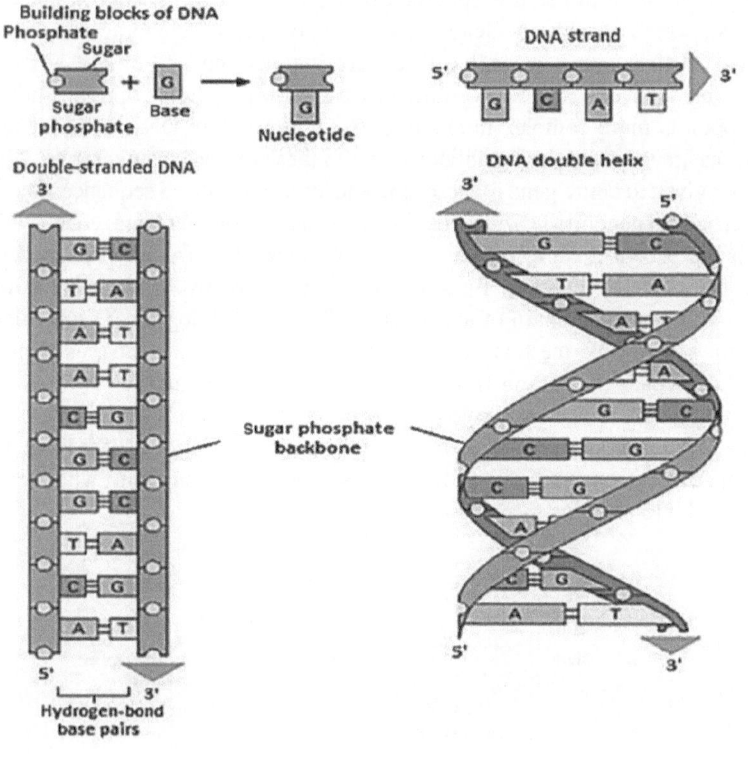

FIGURE 1.2 DNA and its building blocks.

Strand one: 5' – C A T T G C C A G T – 3'
Strand two: 3' – G T A A C G G T C A – 5'

Encoding information to build different proteins is provided by definite continuous sections present along the DNA structure. Based on the type of organisms, these sections represent diverse genes and also vary in length. Locating the stop/start codons and donor/acceptor positions in Figure 1.2 could recognize precisely start and end parts of exon segments and genes within the DNA of eukaryotes. Before synthesis of proteins, conversion to messenger RNA (mRNA) occurs within DNA sequence.

The codon 'AUG' that codes to form methionine and signals the start of a gene is part of this many-to-one mapping [20]. Based on genetic codes, the flow of information outlines relations between the base sequences in the transcript of DNA or its mRNA and amino acids of protein. An amino acid is specified by a sequence of three bases, termed as codon. The block diagram shown in Figure 1.3 explains that the flow of information within DNA results in proteins. In the process of DNA replication or synthesis of proteins, a single exact duplicate copy of a genome is created, prior to the cell division. Each of the two resulting identical double strands is comprised of one newly synthesized strand and one original strand [21]. A cell mechanism (i.e. promoter) recognizes the start point of genes.

Every type of amino acid is specified by triplets formed by mRNA nucleotides, which are termed as codons. Genetic code is known to be mapping of 64 probable codons to 20 amino acids as well as three terminator signals.

DNA then converts into RNA in the process known as transcription. The genetic transcription is more complex in eukaryotes. Only exons in DNA code for proteins and introns are removed during splicing. Due to these regions in eukaryotes, different names are given to entire gene found in chromosome and spliced sequence. The earlier one is termed as genomic DNA, while the latter one as complementary DNA (cDNA).

With the use of an mRNA template, molecules of DNA are produced during reverse transcription process. Finally a translation process translates the mRNA codons into proteins, a chain of amino acids. The central dogma of molecular biology defines the whole method of translation of DNA into proteins, as depicted in Figure 1.4. Typically, moving from DNA to protein in eukaryotes is the so-called central dogma of molecular biology. Transcription remains the first part, where genetic information from DNA is transferred into RNA. Splicing of mRNA is done by eliminating introns and combining exon segments into one unit, which could be transformed into proteins using genetic codes.

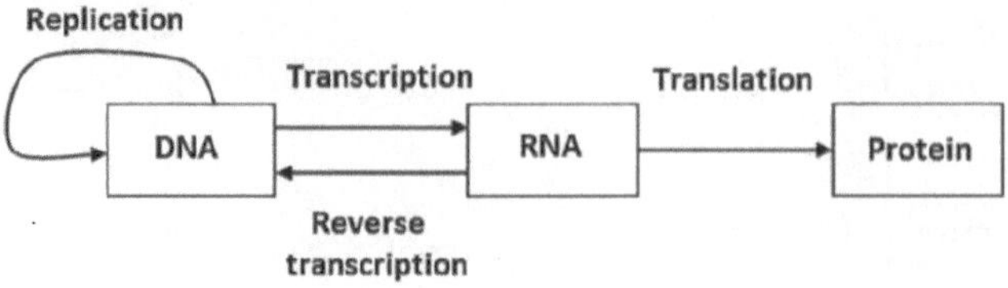

FIGURE 1.3 Genetic information flow in a cell.

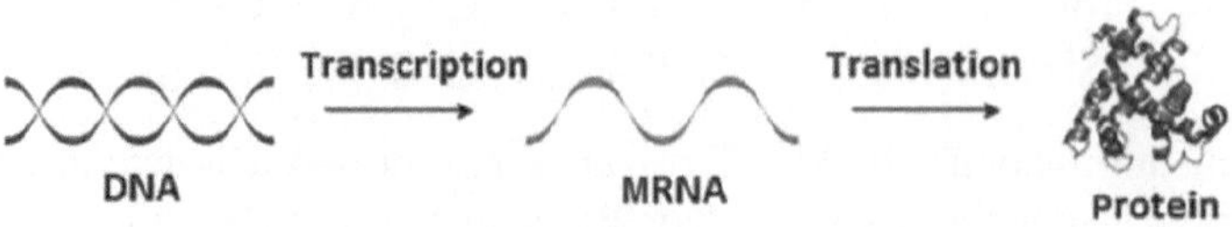

FIGURE 1.4 Central dogma of molecular biology.

1.6 GENE PREDICTION

Recognizing exon segments in computational biology area using algorithms is known as gene prediction. Identification of other DNA functional components such as RNA genes and regulatory networks may also be included.

1.6.1 Significance of Gene Prediction

Gene prediction is the first and best substantial metric for the molecular understanding of the genome after species sequence. Statistical analysis of homologous recombination rates of multiple distinct genes could decide their order on a specific chromosome. Multiple experimental data could be mixed to produce a genetic map that specifies the rough positions of recognized genes compared to one another. Today, with genome sequences of different species available and strong resources for computation available to the research community, gene prediction has largely become a main problem of computation [22].

1.7 TYPES OF GENE PREDICTION APPROACHES

This subsection explains the three types of approaches used for gene prediction, namely, extrinsic, ab initio, and comparative.

1.7.1 Extrinsic Gene Prediction

In the extrinsic approach, an inverse transformation of genetic code is used for deriving probable DNA coding segments. These sequences of candidate DNA are then used to search for matches in a target genome, full or partial, and accurate or inaccurate. The fundamental local alignment search tool BLAST used for this purpose is recognized as the commonly used design of system [23]. A comprehensive sequencing of protein products and mRNA is needed to apply this strategy to gene prediction. In this context, the research of hundreds or thousands of distinct cell kinds introduces additional problems. For instance, some genes of *Homo sapiens* can only be expressed as a fetus during development, which for ethical reasons could be difficult to study.

Regardless of such problems, for human and other model organisms, a number of protein sequence and transcript databases were produced. Examples include the RefSeq database of transcript and protein sequences of many organisms [24], and the Ensembl system with evidence mapping of human and several other genomes [25]. Recently, either an ab initio or intrinsic program for gene identification like AUGUSTUS has been extended using a generalized hidden Markov model (GHMM)-based modeling of extrinsic evidence. Gene identification with a greater accuracy has been reported for the new program known as AUGUSTUS+ [26].

1.7.2 Ab Initio Gene Prediction

Sequences of genomic DNA alone are systematically searched for certain indications of protein coding segments using this technique. These indications may be either certain biological sequences, particular areas indicating the existence of neighboring genes, or the content and statistical characteristics of the sequence of protein coding itself. In prokaryotes, ab initio gene prediction is easy due to relatively well-understood promoter sequences and transcription factor binding sites. The protein coding sequences are fairly long and contiguous. Furthermore, the detection of statistical properties of coding regions is straightforward due to one reasonably long sequence stretch for one gene. GLIMMER is a commonly used and extremely precise prokaryotic gene prediction scheme [27]. In eukaryotes, due to more complex promoter and other regulatory signals, ab initio gene prediction is a challenging problem. There is much more interest for research communities in the RNA processing of eukaryotes than prokaryotes in genomics.

CpG islands, start codons, donor splice sites, acceptor splice sites, stop codon, and poly (A) end need to be detected. Prediction of eukaryotic genes is difficult because of their isolated as well as non-constant nature of genes. In other words, eukaryotic genes are additionally split into comparatively tiny sections of protein coding segments termed as exons, disrupted by introns known as non-coding sections. The intergenic along with intronic areas frequently comprise greater than 95% of their genomes in eukaryotes such as humans. Therefore, statistical characteristics and other known contents of protein-coding DNA in eukaryotes are much harder to identify.

GENSCAN and GeneID are ab initio eukaryotic gene prediction programs. GeneID was one of the first computational gene finding programs depending on a simple hierarchical design [28]. Eukaryotic ab initio gene finders have accomplished restricted success compared to prokaryotic gene prediction. A review of various computational gene finding programs is presented in Ref. [29]. Methods driven by the data from the previous studies of computational gene prediction programs appear towards depending on the statistical data of sequence composition than DNA sequences engaged within the method of shifting as of DNA conversion into protein [30]. The existing programs rely much on the sequence statistics and are less suitable for each sequence. The predictability also relies on the length as well as position of exons. It has been observed that most of the available programs miss completely or partly the prediction of short (<50 base pairs) initial and terminal exons, and techniques have been proposed to address this [31].

Another limitation of the gene prediction programs is that they are very poor at detecting nested or overlapped genes. High predictive accuracy could generally be ascribed to welcoming practice and test sequences following certain guidelines, like sequences comprised of single full gene thru di-nucleotides of intron segments like 'GT' and 'AG,' respectively, for their acceptor and donor splice positions, exclusive of joined gene positions with stop codons existing in the frame. A method to improve the detection of these short coding regions is illustrated in Ref. [32].

On the other hand, it was noted that by combining distinct techniques, accuracy in locating genes can be significantly improved. Therefore, this region needs to be investigated and addressed by examining the function of particular DSP methods, developing different novel DSP methods, and combining better new non-data-driven techniques with current data-driven ab initio gene prediction algorithms to promote the state of the art. Ab initio techniques capable of identifying genes that are not similar to those in databases are vital tools for analyzing meta-genomic sequences [33].

1.7.3 Comparative Gene Prediction

In comparative gene prediction, genomes of associated organisms are compared for resemblance of sequence, gene position, length and amount of gene coding segments or exons, sections related to non-coding DNA, and other extremely preserved segments. For instance, scientists have been experimenting with how human genes compare their counterparts in easier model organisms, for instance, the mouse. Examples of comparative gene prediction programs are EuGene'Hom [34], GPHMM [35], and SLAM [36].

1.8 DNA REPRESENTATIONS FOR GENOMIC SEQUENCE ANALYSIS

Gene finding is a vital problem in the area of bioinformatics. Various bioinformatics approaches and methods based on the Marple algorithm and Wavelet Packets Transform are presented in Ref. [37,38] for gene finding. As a first step, processing DNA sequences using DSP techniques involves their transformation from a string of characters to numerical sequences. Therefore, representing DNA symbols in digital signals remains essential to the gene sequence analysis based on DSP. This section discusses the desirable characteristics for binary notation of DNA followed by a review of various DNA representations in the subsequent section.

1.8.1 Desirable Properties

Numerous approaches are implemented in the recent years for mapping the numeric values of DNA characters (i.e. A, C, G, and T). However, they invariably match the DNA sequences to four numbered signals. Few of the desirable DNA notation characteristics may include the following:

- Similar weight (e.g. size) is exhibited by every nucleotide as no biological proof shows that one is essential more than the other.
- Equal distances should exist between all nucleotide pairs, as there is no biological proof to indicate that any pair is nearer than any other.
- Compact representations should be minimized, in specific redundancy.
- Representations should provide access to a variety of tools for mathematical assessment.

1.9 TYPES OF DNA REPRESENTATIONS

In discussing the properties of each representation, the number of sequences produced by that particular representation and computational complexity of all subsequent processing of numerical sequences are presented in this book. The following section reviews the popular binary Voss DNA representation along with various existing DNA representations.

1.9.1 Voss Mapping

Perhaps the first as well as best famous mapping of DNA is the binary or Voss representation [39]. The groundbreaking part in this system was the fact that a numeric was allocated to every DNA symbol, or a walk displacement along a direction in D-dimensional space with D < 4, introduces numeric correlations. The Voss binary notations of nucleotides A, C, G, and T map the sequences as four binary sequences for nucleotides $x_A[n]$, $x_C[n]$, $x_G[n]$, and $x_T[n]$, which display their occurrence as 1 as well as their nonexistence as 0 at a given point n. The four binary indicator sequences would appear for the brief DNA fragment $x[n]$ = AGTTCTACCGAGC as

$$x_A[n] = \{1,0,0,0,0,0,1,0,0,0,1,0,0\},$$

$$x_C[n] = \{0,0,0,0,1,0,0,1,1,0,0,0,1\},$$

$$x_G[n] = \{0,1,0,0,0,0,0,0,0,1,0,1,0\},$$

$$x_T[n] = \{0,0,1,1,0,1,0,0,0,0,0,0,0\},$$

where $x_A[n] + x_C[n] + x_G[n] + x_T[n] = 1, \forall n$ and n is the index for the base.

A symbolic sequence $s[n]$ comprising M different symbols $S_1,\ldots,S_M$, to which numerical values $a_1,\ldots,a_M$ are assigned, can also be represented as follows:

$$s[n] = a_m \; x_m[n], \; m \in \{A,C,G,T\} \tag{1.1}$$

where $x_m[n]$ are the above-mentioned binary indicator sequences.

These input DNA sequences are obtained from the NCBI genome databank and are converted to binary notation using binary representation [40].

1.9.2 Z-Curve Representation

The Z-curve technique transforms the DNA sequence into a three-dimensional periodic tetrahedrons symmetry equivalent representation [41]. All the three-dimensional *Z* curves for DNA sequences of various species along with their three component curves are stored in the Z-curve database [42]. Originally, the Z-curve has been used to visualize and analyze DNA sequences. Z-curve representation fundamentally produces the same DNA spectrum as that acquired with the Voss representation at a reduced computational cost [43]. It was subsequently used to identify the gene and evaluate DNA protein-coding sections. The DNA sequence is first transformed to four binary indicator sequences, $x_A[n]$, $x_C[n]$, $x_G[n]$, and $x_T[n]$, as calculated in the previous section to produce Z-curve sequences.

The aggregate figures of the DNA bases are then calculated with n specimens using

$$l_n = \sum_{i=0}^{n} x_l[i], \quad l \in \{ A,C,G,T\} \tag{1.2}$$

$$\begin{aligned} x_n &= 2[A_n - A_{n-1} + G_n - G_{n-1}] - 1, \\ y_n &= 2[A_n - A_{n-1} + C_n - C_{n-1}] - 1, \\ z_n &= 2[A_n - A_{n-1} + T_n - T_{n-1}] - 1 \end{aligned} \tag{1.3}$$

The Z-curve sequences can also be calculated using the four binary indicators of the Voss representation as follows:

$$\begin{bmatrix} x_n \\ y_n \\ z_n \end{bmatrix} = 2x \begin{bmatrix} 1 & 0 & 1 & 0 \\ 1 & 1 & 0 & 0 \\ 1 & 0 & 0 & 1 \end{bmatrix} x \begin{bmatrix} x_A[n] \\ x_C[n] \\ x_G[n] \\ x_T[n] \end{bmatrix} - \begin{bmatrix} 1 \\ 1 \\ 1 \end{bmatrix} \tag{1.4}$$

The gene sequence is shown by Z-curve sequences as three signals with values 1 or −1, each of which has a biological interpretation. The relation (1.3) can be used to show that x_n equals 1 for n^{th} nucleotide base G (purine) or A, or −1 for n^{th} nucleotide T (pyrimidine) or C also y_n equals 1 for n^{th} nucleotide base C (amino-type) or A, or −1 for n^{th} nucleotide base T (keto-type) or G; also z_n equals 1 for n^{th} nucleotide base T (weak hydrogen bond) or A, or −1 for n^{th} nucleotide base G (strong hydrogen bond) or C. In addition, the Z-curve technique reduces calculation costs in subsequent processing phases by removing one redundant sequence from the Voss range of four binary indicator sequences. The DNA spectrum based on the Z-curve has been shown to be the same as the one founded by the Voss.

1.9.3 Tetrahedron

The tetrahedron notation too decreases the amount of sequences from 4 to 3, in a way that is symmetrically consistent with entire four components [44]. In this way, the four nucleotides define a path to one corner of the tetrahedron in the representation. A 3D vector may represent each nucleotide (e.g. having *r*, *g*, and *b* coefficients) as follows:

$$\begin{aligned} A &= k \\ C &= -\frac{2\sqrt{2}}{3}i + \frac{\sqrt{6}}{3}j - \frac{1}{3}k \\ G &= -\frac{2\sqrt{2}}{3}i - \frac{\sqrt{6}}{3}j - \frac{1}{3}k \\ T &= -\frac{2\sqrt{2}}{3}i - \frac{1}{3}k \end{aligned} \tag{1.5}$$

A particular advantage of this nucleotide-level representation is that it can be easily extended at the codon and amino acid levels. Three tetrahedral numerical sequences $x_g[n]$, $x_r[n]$, and $x_b[n]$ are calculated using a relation given by

$$x_l[n] = a_l x_{\mathrm{A}}[n] + c_l x_{\mathrm{C}}[n] + g_l x_{\mathrm{G}}[n] + t_l x_{\mathrm{T}}[n] l \in \{r, g, b\} \tag{1.6}$$

whereas $x_{\mathrm{A}}[n]$, $x_{\mathrm{C}}[n]$, $x_{\mathrm{G}}[n]$, and $x_{\mathrm{T}}[n]$ remain four numeric binary sequences, and (a_r, a_g, a_b), (c_r, c_g, c_b), (g_r, g_g, g_b), and (t_r, t_g, t_b) are four 3D vector coefficients. Therefore, the final tetrahedral gene sequences were determined using expressions given by

$$\begin{aligned} x_r[n] &= \frac{\sqrt{2}}{3}(-x_{\mathrm{C}}[n] - x_{\mathrm{G}}[n] + 2 - x_{\mathrm{T}}[n]) \\ x_g[n] &= \frac{\sqrt{6}}{3}(x_{\mathrm{C}}[n] - x_{\mathrm{G}}[n]) \\ x_b[n] &= \frac{1}{3}(3\ x_{\mathrm{A}}[n] - x_{\mathrm{C}}[n] - 2\ x_{\mathrm{G}}[n] - x_{\mathrm{T}}[n]) \end{aligned} \tag{1.7}$$

The representations of Voss and Tetrahedron were shown to be equal for calculation of the power spectrum. Therefore, given the equivalence of *Z*-curve and Voss notation, logically, the spectra of Z-curve and tetrahedron representations are also identical.

1.9.4 Complex

A symbolic representation of DNA sequence using a complex notation is illustrated in Ref. [45]. The complex representation presented is dependent on the notion that four 3D vector coefficients in (1.5) are either +1 or −1 and written as follows:

$$\begin{aligned} \mathrm{A} &= i + j + k \\ \mathrm{C} &= -\ i + j - k \\ \mathrm{G} &= -i - j - k \\ \mathrm{T} &= i + j - k \end{aligned} \tag{1.8}$$

The dimensionality of bipolar form in (1.8) is decreased by placing the fundamental tetrahedron on an appropriate surface, leading in a complex notation of each base of DNA is expressed as follows:

$$\begin{aligned} \mathrm{A} &= 1 + j \\ \mathrm{C} &= -1 + j \\ \mathrm{G} &= -1 - j \\ \mathrm{T} &= 1 - j \end{aligned} \tag{1.9}$$

The complex notation represents few basic properties of mathematical characteristics. Alternative nature of A-T as well as C-G pairs, for example, is expressed through the symmetry of real axis (i.e. complex conjugate representations), and pyrimidine (C or T)/purine (A or G) base pairs present a similar imaginary component. In addition, one complex indicator sequence can be used to represent the DNA character string. Alternatively to typical DNA nucleotide depiction, certain complicated weights can be calculated for each of the four bases.

1.9.5 Quaternion

Quaternions have been known as hyper complex numbers, due to the fact that a quaternion can be represented as $q = q_1 + iq_2 + jq_3 + kq_4$, where i, j, and k are three complex elements (each just like the element i or j in complex numbers). The first element of quaternion, q_1, is referred to as a scalar, and the last three elements q_2, q_3, and q_4 are referred to as the vector parts of the quaternion. A quaternion with all four elements is known as a complex quaternion, whereas a quaternion with last three elements is recognized as a pure quaternion [46,47]. The fundamental formula for quaternion multiplication is

$$i^2 = j_2 = k_2 = i\ j\ k = -1 \tag{1.10}$$

Since multiplication of quaternions is not commutative, the following set of relations can be very easily derived from (1.10):

$$i\ j = k,\ j\ k = i,\ k\ i = j,\ j\ i = -k,\ k\ j = -i, \&\ i\ k = -j \tag{1.11}$$

The conjugate q^* of quaternion $q = q_1 + iq_2 + jq_3 + kq_4$ is $q^* = q_1 - iq_2 - jq_3 - kq_4$. A detailed description of algebra of quaternion was presented. Pure quaternions remain allocated to every symbol in this notation of gene nucleotide bases, as

$$\begin{aligned} &\mathrm{A} = i + j + k, \mathrm{C} = i - j - k \\ &\mathrm{G} = -i - j + k, \mathrm{T} = -i + j - k \end{aligned} \tag{1.12}$$

Each DNA symbol moreover allocated for complex quaternions (i.e. $\mathrm{A} = 1 + i + j + k$, $\mathrm{C} = 1 + i - j - k$, $\mathrm{G} = 1 - i - j + k$, and $\mathrm{T} = 1 - i + j - k$). This system enables to represent gene sequence as 1–4 sequences similar to the complex notation. The subsequent processing of quaternion sequences is, however, limited to specific mathematical tools. In order to identify tandem repeats in the DNA sequences, pure quaternions were used for the calculation of the periodicity transform. A novel method for finding the tandem repeats in DNA sequences is presented using quaternion periodicity transform [48]. In comparison to the normal notation of representing gene nucleotide bases as complex numbers, simple quaternion mapping of DNA symbols offers an improved symbolic sensitivity for transforming to DNA patterns, as well as an uneven selection criterion at threshold. The quaternion strategy can also enhance the identification of DNA patterns in the spectral domain by using the Fourier quaternion.

1.9.6 Electron-Ion Interaction Potential

A nucleotide-specific physical property (e.g. A = 0.1260, G = 0.0806, C = 0.1340, and T = 0.1335) remains useful for mapping the nucleotide bases of DNA to numeric notation, which refers to as electron-ion interaction potential (EIIP) [49]. This strategy also enables for one or four sequence(s) of DNA representation. The informational spectra method (ISM), a physical and mathematical technique of DNA sequence assessment, is the basis for this notation [50]. The following equation has been used to calculate EIIP for each nucleotide:

$$W = \frac{\left(0.252\ \ X\ \ Z^{*}\ X\ \sin\left(1.04X\ \pi\ X\ Z^{*}\right)\right)}{(2\ X\ \pi)} \tag{1.13}$$

where Z^* in (3.13) is a quasi-valence number defined as

$$Z^* = \sum_{i=1}^{m} \frac{n_i\ z_i}{N} \tag{1.14}$$

where Z_i is the valence atomic number of ith component, n_i is the amount of atoms of ith atomic component, m is the amount of components of atom in molecule, and N is the total atoms. This notation has a distinct magnitude for individual base and is unevenly distinguished from each representation.

1.9.7 Inter-Nucleotide Distance

In Ref. [50], each nucleotide is replaced by an amount that denotes the distance from the present nucleotide to the subsequent nucleotide that is similar. The similar value of the present nucleotide is the duration of the rest of the series when the gene sequence is scanned starting from left towards right when the comparable nucleotide base remains not identified.

For instance, for genomic sequence $x[n]$ = AGTTCTACCGAGC, the sequence with inter-nucleotide distance is written as

$$\text{ind}[n] = \{6, 8, 1, 2, 3, 7, 4, 1, 4, 2, 2, 1, 0\} \tag{1.15}$$

From any given ind[n], it is not possible to reproduce the original sequence. However, the decompositions of the ind sequence into two sequences (i.e. paired ind and pairing A-T/C-G) and four sequences do conserve the biological information of the original sequence, although at the expense of extra processing.

1.9.8 Maximum Likelihood Estimate

This strategy assumes that every DNA symbol is produced from a source of data using sequence $D = [D_1, D_2, \ldots, D_N]$ as well as fundamental probability weight function resulted thru drawing the symbols in a cyclic manner [51]. The amount of sources corresponding to the latent period in sequence and the latent periodicity is identical to periodicity in a statistical way. For period P, the maximum likelihood estimate (MLE) is then calculated using the expression

$$\text{MLE} = \arg\max_{p=B} \log p\left(\frac{W}{M^P}, P\right) \tag{1.16}$$

where $B = \{1,\ldots, N_0\}$ is the search space for parameter P ($N_0 < N$), $W = [w_1,\ldots, w_N]$ is the sequence of vectors to represent D, and M is the stochastic matrix whose columns represent the probability mass functions of data sources; entry M_{ji} denotes the probability that i^{th} source generates j^{th} symbol for alphabet $S \in \{A, C, G, T\}$.

1.10 ORGANIZATION OF BOOK

This book addresses the issue of improving the accuracy in exon prediction using various adaptive techniques based on different performance measures, which is crucial in disease diagnosis and therapy. Six chapters are presented in the book. This book is structured and organized as follows in order to achieve the goals of studies:

Chapter 1 presents an overview of genomics engineering, structure of DNA sequence and its building blocks, genetic information flow in a cell, gene prediction along with its significance, and various types of gene prediction methods (i.e. extrinsic, ab initio, and comparative). An account of the objectives of our research work and various types of DNA representations for genomic sequence analysis along with its desirable properties is provided, and the organization of the book is also described.

Chapter 2 presents the review of literature starting with the biological background of genomic sequence analysis followed by the gene and early development of genetics. Moreover, various DSP-based techniques, adaptive algorithms for DNA analysis, and motivations for our research work are discussed, followed by conclusions.

Chapter 3 covers various theoretical considerations of adaptive filtering techniques used for DNA analysis, introduction to adaptive filtering, properties of adaptive algorithms, need for development of AEPs, and the structure of AEP used for DNA analysis in our current work. Also, the familiar LMS algorithm is discussed elaborately. This is considered as the reference algorithm, and various AEP-based realizations developed in our work are compared with this algorithm. Here, we have inspected the possibility to enhance the existing methods in order to achieve better performance. We developed various AEPs using various adaptive algorithms that branch from LMS in the first category. In this chapter, we have presented five types of AEPs developed using LMS, Least Mean Fourth (LMF), Variable Step Size LMS (VSLMS), Least Mean Logarithmic Squares (LMLS), and Least Logarithmic Absolute Difference (LLAD) algorithms. For larger sequences, an adaptive algorithm that offers less computational complexity is desirable.

To achieve this, we have extended the above adaptive algorithms with their sign-based realizations, which are known to reduce computational difficulty. Three simplified versions of algorithms based on signum function are presented in this chapter for this purpose, namely, (i) Signed Regressor Algorithm (SRA), (ii) Sign Algorithm (SA), and (iii) Sign-Sign Algorithm (SSA). Sign-based treatment to all five adaptive algorithms requires appropriate modifications in the parts of weight expression.

Thus, three versions of simplified sign algorithms are combined with all LMS-based algorithms, which results in sign-based versions of that particular algorithm. Versions of LMS such as Sign Regressor LMS (SRLMS), Sign LMS (SLMS), and Sign-Sign LMS (SSLMS) were derived.

Therefore, we have considered five adaptive algorithms here including LMS algorithm, and their sign-based versions result in twenty algorithms for the development of AEPs. This chapter also covers evaluation of the performance of various AEPs using measures like computational complexities, convergence plots, calculation of metrics like sensitivity (Sn), specificity (Sp), and precision (Pr) along with exon prediction results using power spectral density (PSD) and comparing with the existing LMS algorithm. Then, the results and discussion of various sign-based LMS variants are presented in this chapter, which covers various gene datasets from the NCBI database used for gene sequence analysis, and exon prediction results followed by conclusions.

Chapter 4 extends the approach of LMS algorithm and its sign-based realizations with normalization factor for DNA analysis. Generally normalization is employed with LMS-type filters to improve convergence speed. Here, we consider both data and error normalizations. The variant of LMS with normalization of data remains regarded as Normalized LMS (NLMS) technique. In this, the correlation between input reference and error output remains normalized thru square of norm of input reference signal vector. The main benefit with this algorithm is that the size of step can be chosen regardless of signal power handling and quantity of filter coefficients. Extension of NLMS by combining sign algorithms results in Normalized Sign Regressor LMS (NSRLMS), Normalized Sign LMS (NSLMS), and Normalized Sign-Sign LMS (NSSLMS) algorithms. The steady-state error does not rely on the reference input signal power because of the existence of a normalizing factor. Instead of data normalization done in NLMS, error can also be used for normalization. An alternate name for error NLMS technique is Error Nonlinear LMS (ENLMS). Thus, the resultant sign-based realizations considered are ENSRLMS, ENSLMS, and ENSSLMS algorithms.

Similarly, the same strategy is applied to Normalized LMF (NLMF) and Variable Step Size NLMS (VNLMS) techniques. Therefore, we have discussed about various sign versions of NLMS, ENLMS, VNLMS, and NLMF algorithms for the development of AEPs for DNA analysis in this chapter. The resulting signed algorithms of VNLMS and NLMF include VNSRLMS, VNSLMS, VNSSLMS, NSRLMF, NSLMF, and NSSLMF algorithms. Normally in practical situations, when input gene sequence is large, adaptive filter length has to be increased for less computational difficulty. Hence, block processing of input data is used as it considerably decreases the adaptive filter's computing load.

In block processing, the data are handled block by block instead of sample by sample. The maximum normalized version of NLMS is known as Maximum NLMS (MNLMS) algorithm. Several maximum variants of all normalized and error normalized techniques based on signum function are also presented in this chapter for the development of AEPs. Thus, the resulting signed variants for maximum normalized variants of NLMS include MNLMS, MNSRLMS, MNSLMS, and MNSSLMS algorithms are discussed. Similarly, we have developed various AEPs using signed variants and other maximum normalized algorithms including Maximum ENLMS

(MENLMS), Maximum NLMF (MNLMF), and Maximum VNLMS (MVNLMS) algorithms. Therefore, we have developed 32 AEPs using various normalized and maximum normalized based adaptive algorithms for exon prediction in this chapter.

Chapter 5 presents the normalized logarithmic based realizations of LMLS and LLAD adaptive algorithms that include normalized LMLS (NLMLS) algorithm, normalized LLAD (NLLAD) algorithm, and their signed variants. These logarithmic LMS versions based on comparative logarithmic costs are presented in the third category for improving the convergence efficiency as well as stability of an AEP. NLMLS and NLLAD techniques are coupled by algorithms based on sign function for further decreasing the computational difficulty of AEP in real-time applications resulting in NSRLMLS, NSLMLS, NSSLMLS, NSRLLAD, NSLLAD, and NSSLLAD algorithms.

Here, an error-normalized LMLS algorithm in which error is used for normalization instead of data normalization known as Error NLMLS algorithm (ENLMLS) is also presented. Thus, resultant error-normalized sign-based realizations considered are ENSRLMLS, ENSLMLS, ENSSLMLS, ENSRLLAD, ENSLLAD, and ENSSLLAD algorithms. Many AEPs are created and checked with real NCBI genomic samples, and also compared with LMS technique. To considerably lower the computation burden of adaptive filter, block processing of input data is done block by block using maximum normalization.

The maximum normalized versions of NLMLS and NLLAD algorithms are known as Maximum NLMLS (MNLMLS) and Maximum NLLAD (MNLLAD) algorithms. Several maximum variants of all logarithmic-based normalized and error-normalized algorithms based on signum function are also presented in this chapter for the development of AEPs. Thus, the resulting logarithmic-based maximum-normalized signed variants of NLMLS and NLLAD including MNSRLMLS, MNSLMLS, MNSSLMLS, MNSRLLAD, MNSLLAD, and MNSSLLAD algorithms are discussed. Similarly, we have developed various AEPs using signed variants of their logarithmic error-normalized algorithms including Maximum ENLMLS (MENLMLS) and Maximum ENLLAD (MENLLAD) algorithms. Thus, we have developed 32 AEPs using various logarithmic-based normalized and maximum-normalized adaptive techniques for exon identification in this chapter. The performance evaluation is carried with reference to convergence analysis, computational complexities, specificity (Sp), sensitivity (Sn), and precision (Pr), and the results are compared with LMS-based AEP.

Chapter 6 ends this book with an overview of the goals achieved and highlights the primary achievements using all proposed techniques. In the previous chapters, we have developed 84 adaptive exon predictors using MATLAB for genomic sequence analysis. In this chapter, we have presented the summary of various AEPs developed using different adaptive techniques and presented a comparison of all the sign regressor-based AEPs used for accurate exon prediction. Experimental results confirm that all the developed AEPs are superior to LMS in exon prediction applications. Also, as their computational complexity is less when compared to their counterparts, we present that the sign regressor version of all adaptive algorithms in each category is suitable for low-complexity applications in real time. This chapter also discusses recommendations for further studies.

2 Literature Review

2.1 BIOLOGICAL BACKGROUND OF GENOMIC SEQUENCE ANALYSIS

The prominence in molecular biology has changed in the past few years from studying individual genes to exploring an organism's full genome. As the point of ambition and the magnitude of accessible information have increased, the field becomes gradually reliant on computational techniques for gene sequence analysis and protein modeling. The discrete nature of genomic information offers investigation by digital signal processing (DSP) applications in this emerging area. The subsequent sections review the essentials of biology and the background of genomic engineering field, the gene and early development of genetics, and existing DSP and adaptive techniques for DNA analysis.

2.2 THE GENE AND EARLY DEVELOPMENT OF GENETICS

The origin of the term 'gene' is from the Greek word 'genos' meaning 'birth,' 'race,' or 'offspring.' The gene is the fundamental unit that can transmit features from generation to generation. It is comprised of a particular sequence of deoxyribonucleic acid (DNA) or ribonucleic acid (RNA) on a rod-shaped structure known as chromosome occupying a fixed position or locus.

The authors J. D. Watson and F. C. H. Crick (1953) introduced the idea that genes live on chromosomes was first proposed by an American biologist Walter Stanborough Sutton, and Thomas Hunt Morgan received experimental support as presented by the author R. J. Reece later in 1910. Morgan and Columbia University colleagues continued their work and developed various gene mapping techniques. During the period from 1952 to 1966, the genetic code was cracked, and transcription and translation processes were defined. It was indicated that there was no true breakthrough in knowing the molecular nature of the gene, although this classical genetic study is brilliant, until the experiments of Rockefeller Institute Hospital scientists Oswald Avery, Maclyn McCarty, and Colin MacLeord, originated in 1942. For the first moment, they found that genes are made of DNA.

It is followed by the Hershey–Chase tests, a series of studies undertaken by Alfred Hershey and Martha Chase in 1952 which confirmed DNA as the genetic material of all species as described in Ref. [52]. Finding the role of DNA led to contributions of many famous biologists during a second great age of genetics.

Since the beginning of the Human Genome Program (HGP) in 1990, the human DNA and model organism gene databanks have grown rapidly. Computational gene forecast becomes increasingly important for automated analysis as well as annotations of big uncharacterized genomic sequences. A lot of gene forecast programs have been created over the past two centuries. The higher gene densities typically associated with prokaryotes and the absence of introns in its exon regions make it less difficult to determine genes in prokaryotic genomes.

Proteins encoding DNA sequences remain transcribed to mRNA without substantial modification; mRNAs are generally converted into proteins. In the same reading frame, lengthy ORFs (open reading frames) ranging from the mRNA initially available to the next stop codon usually give a good, but unsure, prediction to protein-coding regions. Various techniques have been used to determine the differences between the compositions of the coding areas, the shadow code areas (coding on the reverse strand of DNA), and non-coding areas using separate types of the Markov models. The majority of the protein-coding genes with outstanding outcomes appear to be defined by such methods as the commonly used GENMARK and Glimmer program. The DNA sequences should be transformed into numeric sequences in order to use DSP principles to evaluate DNA sequences. The DSP methods are therefore used in the study of DNA.

The DSP methods used for DNA assessment are Discrete Fourier Transform (DFT), Discrete Wavelet Transformation (DWT), Digital Filtering, Parametric Modeling, and Entropy. The major property of three-base periodicity is exhibited by the DNA coding sections. In Ref. [53], the authors Shrish Tiwari et al. (1997) presented Fourier's techniques for analyzing this regularity and recognizing the coding segments present in the DNA of genome. The three-base periodicity in the structure of nucleotides is shown in the Fourier spectrum as a sharp peak at $f = 1/3$ frequency. In Ref. [54], the authors Salzberg S. L. et al. (1999) presented a novel scheme, GlimmerM, designed to discover genes in the malaria parasite, *Plasmodium falciparum*. The scheme was expanded to discover splice sites with specially trained modules on all available *P. falciparum* genome information. Because gene density of *P. falciparum* is comparatively large, a good bacterial gene finder, the Glimmer, was the basis for design of the system. Although an accurate assessment of its precision is difficult at this moment, all these predictions were verified by laboratory trials on a tiny variety of predicted genes. The accessibility of this new gene finder will significantly facilitate the annotation process with the fast advancement in sequencing the genome of *P. falciparum*.

The authors Morgenstern B. et al. (2002) in Ref. [55] outlined a novel application of the DIALIGN sequence-alignment program created to align big genomic sequences. They likened this technique to the PipMaker, WABA, and BLAST alignment programs and demonstrated that local similarities recognized by these programs are extremely associated with protein-coding areas. The authors have found that the relatively highest maximum height within the Fourier spectrum at $f = 1/3$ is more likely to discriminate against coding potential, based on an extensive spectral analysis of full-length DNAs over 5.5 million base pairs from a wide variety of organisms and by a separate examination of coding and non-coding gene sequences.

The local signal-to-noise proportion of the maximum is examined in a sliding window to detect possible exon regions in gene sequences. In 2002, the authors P. P. Vaidyanathan et al. (2002) in Ref. [56] provided effective digital filters for detection of the period-3 sections, helping to predict gene sites and allowing the prediction of particular exons within eukaryotic cell genes. Digital filters were shown not only to obtain the period-3 element but also to effectively eliminate the background $1/f$ spectrum displayed in almost every DNA sequence.

The authors Daniel Kotlar et al. (2003) in Ref. [57] have presented a new measure by name Spectral Rotation Measure to identify the DNA sequence protein-coding areas. This measure is based on a 1/3 frequency phase of the DFT, calculated with the four binary sequences A, C, T, and G. A number of metrics depending on the phase property are proposed. The measurements are calculated by rotation of the vectors in clockwise direction, achieved thru DFT on every frame of evaluation, at an angle similar to the respective core value. All vectors of the complex plane in protein-coding areas are presumed to be aligned closely, amplifying the vector sum magnitude. Magnitude does not considerably alter here in non-coding areas. The phase property is also used to find the reading frame of the sequence. Digital filtering and DFT relevant methods used to identify coding areas do not substantially suppress the DNA spectrum of non-coding regions at $2\pi/3$. Non-coding segments may therefore be recognized as exon segments inadvertently. The authors Trevor W. Fox et al. (2004) described in Ref. [58] a novel method that eliminates almost all non-coding areas and enhances the probability of properly recognizing coding regions in such genes.

The authors David DeCaprio et al. (2007) in Ref. [59] have presented Conrad, the first semi-Markov conditional random field (SMCRF) comparison gene predictor. Conrad, as opposed to the best independent genetic predictors, is trained on the basis of generalized hidden Markov models (GHMMs) and is trained by the highest probability, in order to maximize annotation precision. In addition, Conrad encrypts all information sources as a feature and handles all features of learning and inference procedures equally as opposed to the finest annotation pipelines which rely on heuristic and arbitrary decision rules to mix standalone genome predictors with additional information such as Expressed Sequence Tag (EST) and protein homology.

In addition, Conrad encrypts all information sources as a feature and handles all features of learning and inference procedures equally as opposed to the finest annotation pipelines which rely on heuristic and arbitrary decision rules to mix standalone genome predictors with additional information such as ESTs and protein homology. Because of its extremely modular nature, SMCRFs are a likely framework for gene prediction, which simplifies the design method and tests of potential indicators of genetic composition. Conrad's SMCRFs advance the state-of-the-art gene identification in plants and provide a strong base for the current and new studies. The writer Jeremy M. Berg in 2008 also illustrated that they had succeeded in producing an extensive assessment of the comparative positions of more than 2000 genes on four chromosomes of fruit fly, *Drosophila melanogaster*, by 1922.

Digital filters were used to predict genes and proteins, but when the characteristic frequency or periodic behavior is altered, the filters need to be redesigned. Later the author Baoshan Ma in 2010 presented a novel approach based on adaptive filtering theory using LMS adaptive algorithm which can identify genes or proteins in a genomic sequence. The authors Hamidreza Saberkari et al. (2013) presented a novel rapid algorithm in Ref. [60] to explore the place of exons in DNA strand based on the mixture of the Goertzel method and Linear Predictive Coding Model (LPCM). The authors Inbamalar T. M. et al. (2013) in Ref. [61] presented new methods to analyze the DNA sequences. This algorithm increases the speed of the process and reduces computational complexity.

It is a quite distinct issue in eukaryotic organisms which was found within prokaryotes. Transcription of exon segments launched in particular promoter sequences is accompanied by elimination of non-coding sequences from pre-mRNA thru a splicing mechanism, thus resulting in exon regions. The subsequent functional mRNA can be converted from the first starting codon to the first stop codon, once introns are withdrawn and some other changes are rendered to mature RNA, usually in the 5' to 3' direction. The encoding of the ORF gene will disrupt the presence of introns that generate stop codons due to intron regions within the eukaryotic gene sequences. This study concentrates on the more complex problem in eukaryotic models of gene identification.

2.3 ORIGIN OF THREE-BASE PERIODICITIES IN GENOMIC SEQUENCES

In genomic sequences, periods of 3, 10.5, 200, and 400 were recorded [62]. Analysis of 10–11 base pair (bp) oscillations in complete genomes of yeast, archaea, bacteria, and their two possible origins is presented in Ref. [63]. The peaks beyond 35 bp mainly reflect the DNA curvature. The measure Counting Oligomers presented in Ref. [64] is more effective than any of the more sophisticated measures. Different measures contain different information. The significance of measuring distinct periods for a specified DNA in locating exon sections, and these periods have been used as discriminating characteristics in several gene prediction research.

The invariant Fourier transform is used to define labeling of a sequence of bases along a given stretch of DNA and therefore can be used as a measure of periodicity for DNA sections having differing base contents [65]. Short-range correlations in nucleotide arrangements, in particular a three periodicity resulting from the repetition of the same nucleotides in the same triplet (codon) places in these regions, are a distinctive feature of genomic exon areas. Periodic sequence A– – A – – A – – A – – ..., where one of four nucleotide bases A, C, T, or G can fill blanks ' – ' randomly.

This sequence provides three periodicity owing to base A repeat at place one of each codon. Periodicity itself is the consequence of certain codon combinations with various typical species frequencies. Instead, the duration of time equal to three is triggered by the genetic code's triple nature. An easy technique for calculating the distance between the comparable nucleotide frequencies is to discover the existence of a periodicity of three. Distance values of 2, 5, 8, and so on can be found in the sequence for a period of three between a specific nucleotide at one position and other similar nucleotides. Stephen T. Eskesen et al. (2004) in Ref. [66] have investigated the cause of DNA periodicity by comparing real and simulated gene sequences. They suggest that codon usage frequencies determine the periodicity of DNA in exons.

Furthermore, the periodicity of DNA itself and the duration of the period equal to three are different. The former is the consequence of certain combinations of codons with distinct typical species frequencies, whereas the latter is triggered by the genetic code's triple nature. Periodicity-3 property of exons was usually defined using methods based on DSP after DNA transformation to numerical notation.

2.4 DSP-BASED TECHNIQUES FOR DNA ANALYSIS

The gene forecast accuracy remains constrained, despite the existence of different data-driven programs for locating genes. One way to increase this accuracy is to combine different types of gene-finding approaches (i.e. extrinsic, ab initio, and comparative) in one program, similar to AUGUSTUS. Alternatively, higher accuracy can also be achieved by combining different gene-finding programs. In both situations, the data dependence of the resultant system would increase. We believe that the combination of improved DSP techniques with other methods driven thru data can also progress thru advancement in locating genes, and the resulting scheme could provide a significantly greater amount of gene prediction precision than other approaches. DSP-based techniques are appealing because, unlike current data-driven approaches, they mostly do not require genomic data training.

They are also obtained from these methods from distinct data. The distinct nature of genomic information that is discrete in both 'time' and 'amplitude' invites DSP methods to investigate. The transformation of nucleotide alphabetic symbols to a number of digital values allows for new as well as helpful applications of DSP to solve issues linked with sequence analysis, such as discovering genes in eukaryotes. The author D. Anastassiou has presented some of the DSP applications and some of the key issues and challenges in the evolving genomic signal processing area in his work. The possible role of DSP applications in ab initio eukaryotic gene finding is summarized in Figure 2.1.

DSP techniques are applied on DNA sequences that have been converted to extract characteristics into numerical signals, which are one dimensional as in virtually all existing DSP-based approaches. The protein-coding/non-coding classification can be achieved using an empirically selected decision threshold for one-dimensional features, or pattern recognition tools (e.g. GMM, SVM) for multi-dimensional features. Finally, DSP-based classification results can be combined with other approaches to provide precise identification of exonic endpoint signals (acceptor/donor splice sites). Firstly, DNA sequences are transformed into appropriate numerical values for the application of DSP techniques. In the past, many DNA representations such as Voss, Z-curve, tetrahedron, complex, quaternion, electron-ion interaction potentials (EIIP), real-numbers, and inter-nucleotide distance have been introduced.

Each of these offers different properties; however, they invariably map the sequences of DNA between one and four digital sequences. For the four nucleotides of gene sequence, various representations set out values imposed in the actual DNA sequence on non-present mathematical structures. For exon forecast relevant to DSP, structure differences of exon as well as intron sections could not be fully exploited thru current mapping systems of genome. For example, the information that occurrence of particular DNA nucleotides in coding sections is different, as for non-coding areas, can be incorporated in representations to be used for the DSP-based gene finding.

Methods for periodicity detection such as DFT, autocorrelation function (ACF), autoregressive (AR) models, and digital filters are then applied to numerical sequences to identify genomic exon regions [67]. Low gene and exon prediction accuracy of the existing DSP-based methods is mainly due to their sole reliance on period-3 identification. Furthermore, these techniques are not ready for capturing supplementary characteristics relevant to genomic regions, and they also address background noise within nucleotide exon identification levels. Most of the existing period-3 detection methods use sliding window attributes to analyze DNA sequences with window length N and the amount of shift between two consecutive windows as one nucleotide to preserve a decent base domain resolution.

The author D. Anastassiou has presented DFT-based methods in which the 351 window size argument was used for making the information window reasonably long or for a few hundred base pairs length. The rectangular window has been widely used in detection of period-3 based on DFT methods. The authors Suprakash Datta et al. (2005) in Ref. [68] have presented an enhanced period-3 detection based on DFT with use of information window of Bartlett as compared to that of rectangular window. In AR modeling, the calculation of its model needs comparatively some sample values, which is useful when the exon areas are brief and/or close to each other. To minimize the spectral leakage in locating exon areas based on DFT, a wider window size is required, meaning longer computational times as well as base domain resolution is compromised.

Digital filters for period-3 detection somehow address this issue and offer a comparable precision of locating exon sections with use of DFT. With use of exon-level identification, a feature value higher than a choice limit within exon for every nucleotide remains adequate to detect that specific exon. By considering the characteristic nucleotide values within the exon, a particular method of detection can give more insight into the robustness. For all existing gene and exon prediction methods, using a sum-of-squares method, characteristics are typically combined to create a one-dimensional function to compare some predetermined boundary. As the technique of sum-of-squares reduces the dimensionality of the features and does not necessarily result in optimal fusion of features, multi-dimensional gene extraction, and exon prediction features.

The resulting characteristics are then passed on to classification between protein-coding and non-coding areas for back-end processing. For one-dimensional classification, a decision threshold derived empirically can be applied, whereas the use of well-known tools for pattern detection such as Gaussian (GMM) or matrix support devices (SVMs) for multi-dimensional evaluation can be achieved. The DSP-based work can be combined with statistical data-driven methods for a more accurate recognition of biological signals at end points. After converting DNA symbols into the numerical sequences using DNA binary representation, some of the existing methods dependent on DSP meant for periodicity identification remain useful for these sequences to acquire one- or multi-dimensional gene, and characteristics of exon prediction are discussed in the following sections.

2.4.1 Application of Discrete Fourier Transform

The DFT was the most common technique used to analyze the spectrum in the finite element series $x[n]$ of the length N is defined as follows:

$$X[k] = \sum_{n=0}^{N-1} x[n] e^{-j2\pi nk/N}, \quad 0 \le k \le N-1 \tag{2.1}$$

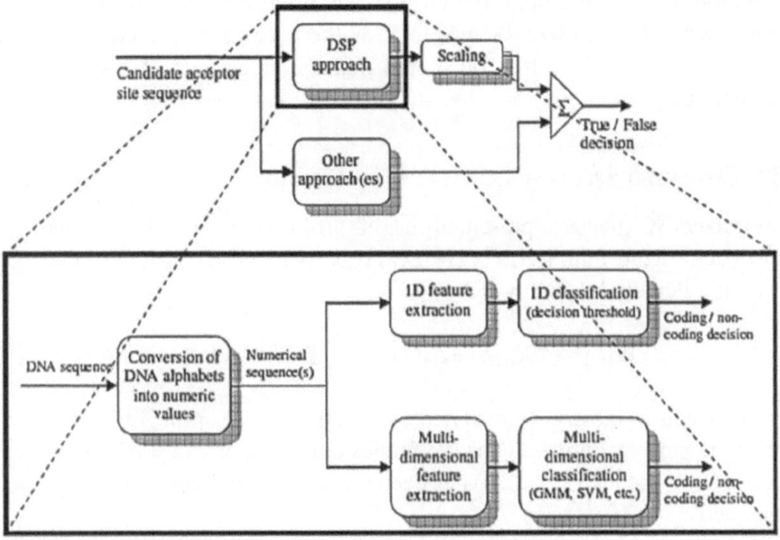

FIGURE 2.1 Existing DSP techniques applied on DNA sequences [67].

where $X[k]$ is the finite-length frequency domain sequence of length N.

It is clear from $x[k]$ which involves N^2 multiplications and $N(N-1)$ additions with direct DFT-computation which is costly for large N_{value}. In order to facilitate DFT applications for large sequences, many efficient methods have been developed for fast DFT computing. Among these and other available methods, the Cooley–Tukey FFT algorithm is the most popular FFT algorithm, and it greatly extends the DFT applications for large sequences by reducing the time complexity ranging from $O(N_2)$ to $O\left(N\log_2 N\right)$. Equation for $X[k]$ of DFT in (2.1) can be used to calculate the DFTs (i.e. $X_A[k]$, $X_C[k]$, $X_T[k]$, $X_G[k]$) of four numeric sequences using a sliding DFT window.

The three base periodicity property shows that in each DFT sequence, the DFT coefficient of $k = N/3$ corresponds to the DNA sequence in the exon regions. Note that calculation of DFT on this single frequency is adequate to allow use of the Goertzel algorithm in Ref. [69], which lowers the cost of DFT computing with almost a factor of two in single point. Based on evaluation using performance metrics such as sensitivity, specificity, and correlation coefficient, this algorithm performs better than the conventional DFT method. Several DFT based spectral measures are provided that uses the three base periodicity behavior to identify the exon sections.

2.4.2 Spectral Content (SC) Measure

The SC measure combines different DFTs (i.e. $X_A[k]$, $X_C[k]$, $X_G[k]$, $X_T[k]$) to acquire a complete DNA spectrum of the Fourier magnitude, as follows:

$$S[k] = \sum_m \left|X_m[k]\right|^2, m \in \left\{\text{A,C,G,T}\right\} \tag{2.2}$$

Based on the SC measurement, the GENSCAN program calculates the peak SNR at $k = N/3$ with $P = S\left[N/3\right]/S'$ where S' is the average of the complete Fourier spectrum (over k) in (2.2). It is presumed that the areas with $P \geq 4$ are protein-coding regions.

2.4.3 Optimized Spectral Content (SC) Measure

The optimized SC measure presented in Ref. [70] assigns complex weights a, c, g, and t to each of the four DFTs $X_A[k], X_C[k], X_G[k]$, and $X_T[k]$. The final Fourier spectrum is calculated as follows:

$$W[k] = \left|a.X_A[k] + c.X_A[k] + g.X_A[k] + t.X_A[k]\right| \tag{2.3}$$

The complicated weights are calculated using a method of optimization applied to a certain organism's recognized genes. Complex conjugate pairs $t = a^*$ and $g = c^*$ can also be applied. The metric in (2.3) has been shown to give a significant improvement over the metric in (2.2).

2.4.4 Spectral Rotation (SR) Measure

A change relevant to the SC measure was proposed by the authors Kotlar and Lavner in Ref. [71]. They discovered DFT distributions of stage angle for exons at frequency $2\pi/3$ are smaller around a key value compared to almost uniform non-coding sections within one specific organism's genomic sequences. They proposed the SR measure using four DFT based rotating vectors $X_A[k]$, $X_C[k]$, $X_G[k]$, and $X_T[k]$ in clockwise direction, each one with an angle equivalent to the average value of stage angle within coding segments σ_m, for making them at every point the same way, as explained in Figure 2.2.

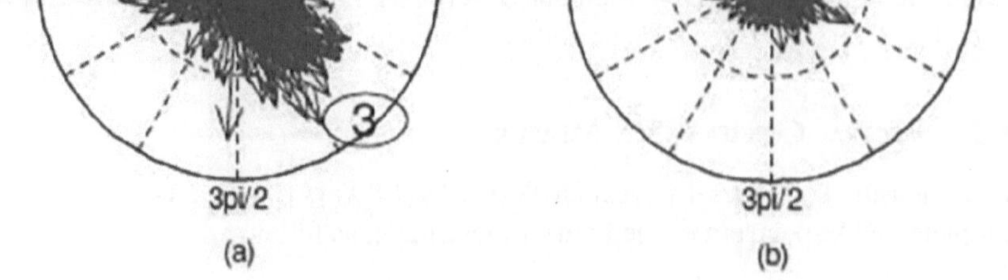

FIGURE 2.2 Output vectors of spectral rotation used by the GENSCAN test set for (a) coding and (b) non-coding regions [71].

The subsequent vectors of coding segments can be seen to be relatively large and 'point' mainly in one of the three feasible paths, whereas vectors of intron regions are tiny in magnitude and pointing in random directions. The SR measure also splits each term in order to give more weight to smaller distributions by the respective phase angle deviations σ_m.

The feature

$$\mathrm{SR}[k] = \left| \sum_m \frac{e^{-j\mu_m}}{\sigma_m} \right|^2, \; m \in \mathrm{A, C, G, T} \tag{2.4}$$

for exon detection function has been used. The SR measure showed a better efficiency than the SC (2.2) and optimized SC (2.3) measurements at a false-positive gene detection level of 10%. Apart from the DFT-based SC, optimized SC and SR measures, the other less popular DFT-based algorithms are combined GeneScan and length shuffle. The authors Datta and Asif in Ref. [72] also showed a promising gene-level identification using a fast DFT-based algorithm.

2.4.5 Fourier Product Spectrum (FPS) Method

Fourier product spectrum (FPS) method has been used for identification of weak estimate signals relevant to tandem repeats hidden within DNA sequences. An estimated tandem repeat is a neighboring section of DNA consisting of one specific pattern that repeats itself one by one imperfectly [73]. It is claimed that in the sum spectrum approach, any binary indicator sequences aperiodic with period p act as additive noise and may result in an unclear peak at frequency $\theta = 1/p$ as presented in Ref. [74]. This limitation is avoided in the FPS which multiplies the individual DFTs (i.e. $X_{\mathrm{A}}[k]$, $X_{\mathrm{C}}[k]$, $X_{\mathrm{G}}[k]$, $X_{\mathrm{T}}[k]$) as below:

$$\mathrm{FPS}[k] = \prod_m \left| X_m[k] \right|, m \in \{\mathrm{A, C, G, T}\} \tag{2.5}$$

2.4.6 Digital Filters for Genomic Analysis

To reduce the spectral leakage of the exon forecast dependent on DFT, it is necessary to have a larger window, which means longer computational time as well as compromises the resolution of base domain. Anti-notch (AN) filter strategy to infinite impulse response (IIR) described in Ref. [74] tries to resolve these issues. The AN filter's response of magnitude shows a sharp peak at $\theta = 2\pi/3$. Numeric sequences are carried through such a filter to detect the period-3 behavior. The magnitude of the output is expected to be larger for exonic than for intergenic and intron regions. Here, the binary indicator sequences are passed through the IIR AN filter. The filter output-based measure gives peaks for exonic regions.

IIR filters require much shorter orders and are more suitable for this application than their finite impulse response (FIR) counterparts. All pass filters can be constructed from the second order. The transfer function of such a real coefficient filter with poles at $Re^{\pm j}$ is given by

$$A(z) = \frac{\left(R^2 - 2R\cos\theta z^{-1} + z^{-2}\right)}{\left(1 - 2R\cos\theta z^{-1} + R^2 z^{-2}\right)} \tag{2.6}$$

where R is the two pole radii in the z-plane, and $R^2 < 1$ for stability. It was shown that expression (2.6) can be used to design a filter having the AN (i.e. complement of a notch) property, if tuned at $\theta = 2\pi/3$.

The following expression is presented to define the AN filter as follows:

$$H(z) = \frac{(1 - A(z))}{2} \tag{2.7}$$

The frequency response of the AN filter defined in (2.7) can be sharp arbitrarily by selecting R near to unity. However, an R value that remains too closer to unity will make effects relevant to round off noise noticeable and the important portion of the filter's impulse response is also going to be very long and compromise the base domain resolution of exon identification. Leave output of the AN filter with an indicator sequence $x_m[n]$ to be input, let AN filter output in (2.7) be denoted by $y_m[n]$, where $m \in \{A, G, C, T\}$.

The respective inputs of the four binary sequences can be combined to create a digital filter-based metric for gene and exon prediction is expressed as

$$Y[n] = \sum_m \left|y_m[n]\right|^2 \tag{2.8}$$

Digital period-3-based filter detection findings have been shown to be similar to the DFT-based SC measure provided by (2.2). A multistage narrowband band pass filter has also been used to improve the prediction accuracy of IIR AN filter. This multistage design cascades $H_1(z^3)$, a multiband version of the low-pass prototype filter $H_I(z)$, with another $H_2(z)$ filter, filter that significantly dampens the zero-frequency pass band with the expression given by

$$H(z) = H_1\left(z^3\right).\ H_2(z) \tag{2.9}$$

whereas $H_I(z)$ is implemented using an all pass filter of the first order $A_0(z)$ and the second order all pass filter $A_I(z)$, both with real coefficients, and $H_2(z)$ remains selected to have two zeros at $z = 0$, i.e. given by

$$H_1(z) = \frac{A_0(z) + A_1(z))}{2} \tag{2.10}$$

$$H_2(z) = \left(1 - z^{-1}\right)^2 \tag{2.11}$$

The resultant narrow band filter $H(z)$ in (2.9) is shown to predict exons with good accuracy, when one of the pass bands of $H_1(z^3)$ is centered at $2\pi/3$.

2.4.7 Autoregressive Models

Autoregressive (AR) methods are an alternative to calculating the signal spectrum. AR models can be described in terms of an all-pole digital filter:

$$H(z) = \frac{1}{1 + \sum_{i=1}^{p} a_l z^{-1}} \tag{2.12}$$

where a_l are the coefficients of the order p model. The following expression of the power spectrum estimation provided in equation for the AR model in (2.11) may be used to detect exon areas specified as

$$\hat{p}(k) = \frac{\sigma^2}{\left|1 + \sum_{l=1}^{p} a_l\, e^{-j\left(\frac{2\pi kl}{N}\right)}\right|^2} \tag{2.13}$$

where σ^2 is the input variance and k represents the power spectrum points. AR technique has a special benefit that calculation of the AR model requires relatively few base pairs, and remains convenient for closer and shorter length exon areas. Also, resolution of detection using AR method remains higher than that of Fourier methods for small DNA sequences [75].

Spurious spectral peaks of high-order AR models limit the efficacy of spectrum-based detection. N. Chakravarthy et al. (2004) in Ref. [76] presented the use of AR modeling for residual error and feature-based analyses of DNA sequences. Note that AR models could not be implemented fairly to numeric notation since they could not be the outcome of an AR method. Before their implementation to AR modeling, the band pass filtering of binary sequences is a possible solution to the problem.

2.5 ADAPTIVE ALGORITHMS FOR DNA ANALYSIS

Precise prediction of gene locations in a genomic arrangement remains a key as well as interesting task in bioinformatics area. Also, precisely forecasting the gene positions remains an essential and difficult job. It is one of the first and vital measures in knowing molecular nature of the genome after sequencing the genome of a specific species. Adaptive algorithms perform a major part in DNA analysis in the area of bioinformatics. A basic adaptive method is the Least Mean Squares (LMS) algorithm. Because of its ease in application, this algorithm is commonly used. Depending on input signal, the adaptive filter may vary its coefficients of weight.

Adaptation occurs through the modification of the parameters of the filter based on information of the incoming signal so that a given performance index is optimized. Therefore, there is almost no previous knowledge of signal as well as noise characteristics. Since no or almost no prior information is available, an adaptive filter needs adaptation or an initial learning period. During this period, its performance is unsatisfactory. The filter should act optimally when tracking non-stationary signals after initial adaptation. In our current work, AEP developed with use of existing LMS adaptive method is also considered in comparison with all the developed AEPs in terms of different performance measures for DNA analysis.

2.6 CONCLUSIONS

In this chapter, the significant developments for exon prediction in the field of bioinformatics from its origin are reviewed covering the biological background of genomic sequence analysis followed by gene and early development of genetics. Then, the origin of three-base periodicities in genomic sequences is also discussed. Several types of gene prediction approaches, DSP-based techniques used for DNA analysis that include DFT, spectral content measure, optimized SC measure, spectral rotation measure, FPS method, digital filters, and autoregressive models from the literature are discussed. Different DSP-based methods and the existing LMS adaptive algorithm presented are less efficient in terms of accurate exon prediction. Finally, an account of existing adaptive techniques for genomic analysis followed by motivation of our research work is discussed.

3 Sign LMS Based Realization of Adaptive Filtering Techniques for Exon Prediction

3.1 INTRODUCTION

In the field of bioinformatics, precise identification of a gene in a genomic sequence is crucial and demanding. The first and most significant step is to know the molecular aspects of DNA followed by gene sequencing of a species. Adaptive algorithms can handle very lengthy sequence strands in various iterations in the encoding of the genomic sequence and can alter weight values according to its statistical nature. A basic adaptive method is the least mean square (LMS) algorithm which is also used for exon prediction [77]. This algorithm is commonly used due to its ease of application. The adaptive filter is able to alter its coefficients of weight based on the incoming data. Adaptation occurs through the modification of the parameters of the filter based on the information of the incoming signal so that a given performance index is optimized. They therefore involve little or no previous understanding of the features of the signal or noise. Since no or almost no prior information is available, the initial learning period or adaptation requires an adaptive filter. During this period, its performance is unsatisfactory. The filter should operate optimally during the tracking of non-stationary input modifications after initial adjustment.

Adaptive LMS-based algorithms make use of the factor that error is a filter weight vector function. Therefore, to identify the state of adaptive process, a single value for objective function occurs for each tap weight vector. The idea remains to iteratively update the tap weights in a way that every new instance exists within the direction of gradient. In this chapter, we develop sign LMS-based adaptive methods, in which the weight update equation is varied with respect to that of LMS. The fundamentals of these methods are presented in Ref. [78]. In addition to the exon prediction accuracy, computational difficulty, convergence efficiency, and prediction accuracy, the ultimate aim is to provide a comprehensive overview of multiple AEPs produced using LMS-based adaptive realizations.

Specificity (Sp), sensitivity (Sn), and precision (Pr) are the measures for performance of exon prediction considered in the present work. Here, we took into account four methods obtained from the LMS method, namely, least mean fourth (LMF), variable step size LMS (VSLMS), least mean logarithmic squares (LMLS), and least logarithmic absolute difference (LLAD). All these LMS-based adaptive algorithms are coupled with sign algorithms to further reduce computational difficulty and improve convergence efficiency. Table 3.1 presents an overview of all the algorithms employed in this chapter.

3.2 THEORETICAL CONSIDERATIONS OF ADAPTIVE FILTERING TECHNIQUES IN DNA ANALYSIS

The word 'filter' is frequently used to describe any system or device that takes a combination of particles or components as its input and produces a matching number of elements or components as its output by processing them based on particular regulations. In particular, a filter is a system that reshapes input frequency elements to produce an output signal with some desirable characteristics. An overall schematic diagram of the filter is presented in Figure 3.1 that stresses its function in restructuring input towards matching desired signal. In this section, an important and elaborated theory related to adaptive filtering, explaining the adaptive filter, properties of adaptive algorithms, and the need for development of adaptive exon predictors (AEPs) is discussed.

3.2.1 Adaptive Filter

In practical situations, most signals are non-stationary in nature; however, the standard finite impulse response (FIR), infinite impulse response (IIR), and notch filters have all defined coefficients of filter and apply the same degree of signal conditioning irrespective of the degree of noise on input signal. In such situations, these fixed coefficient, non-adaptive filters are not suitable. In order to filter time-varying non-stationary components, the filter coefficients must vary in accordance with the features of input signal. These adaptive filters are able to alter the filter coefficients according to the characteristics of the input signal.

3.2.2 Properties of Adaptive Algorithms

The fundamental operation of an adaptive filter reduces an objective function, generally denoted by J_w. By a proper selection of this objective function, one can condition it to have a continuous shape. One or more of the aforementioned factors determines the decision of one algorithm over another.

1. *Convergence Rate:* It is defined as the quantity of iterations required to complete "near enough" in mean square error (MSE) response to stationary inputs to an optimal wiener response. The convergence curve can be plotted by measuring MSE in each iteration and plotting the curve among a number of iterations and MSE. The relation of MSE is

$$J = E\left\{|e(n)|^2\right\} \tag{3.1}$$

where J is the MSE and $e(n)$ is the error signal.

TABLE 3.1
Summary of Algorithms Used in Chapter 3

S. No.	Acronym	Name	Advantage
1	LMS	Least mean squares	Weight drift problem can be avoided and stability increases
2	SRLMS	Sign regressor least mean squares	Single multiplication required and data is clipped
3	SLMS	Sign least mean squares	Error is clipped
4	SSLMS	Sign regressor least mean squares	Zero multiplications; both data and error are clipped
5	LMF	Least mean fourth	Lower steady-state error than LMS
6	SRLMF	Sign regressor least mean fourth	Single multiplication required; data is clipped; lower steady-state error than LMS
7	SLMF	Sign least mean fourth	Error is clipped; lower steady-state error than LMS
8	SSLMF	Sign-sign least mean fourth	Zero multiplications; both data and error are clipped; lower steady-state error than LMS
9	VSLMS	Variable step size least mean squares	Lower steady-state error than LMS
10	SRVSLMS	Sign regressor variable step size least mean squares	Single multiplication required; data is clipped; lower steady-state error than LMS
11	SVSLMS	Sign variable step size least mean squares	Zero multiplications; both data and error are clipped; lower steady-state error than LMS
12	SSVSLMS	Sign-sign variable step size least mean squares	Zero multiplications; both data and error are clipped; convergence performance; lower steady-state error than LMS
13	LMLS	Least mean logarithmic squares	Weight drift problem can be avoided and stability increases
14	SRLMLS	Sign regressor least mean logarithmic squares	Single multiplication required; data is clipped; weight drift problem can be avoided; stability increases
15	SLMLS	Sign least mean logarithmic squares	Error is clipped; weight drift problem can be avoided; stability increases
16	SSLMLS	Sign-sign least mean logarithmic squares	Zero multiplications; both data and error are clipped; weight drift problem can be avoided; stability increases
17	LLAD	Least logarithmic absolute difference	Uses logarithmic cost function with normalized error; faster convergence than LMLS and other variants
18	SRLLAD	Sign regressor least logarithmic absolute difference	Less computational complexity than LLAD and other variants; faster convergence than LMLS algorithm
19	SLLAD	Sign least logarithmic absolute difference	Error is clipped; weight drift problem can be avoided; stability increases
20	SSLLAD	Sign-sign least logarithmic absolute difference	Zero multiplications; both data and error are clipped; weight drift problem can be avoided; stability increases

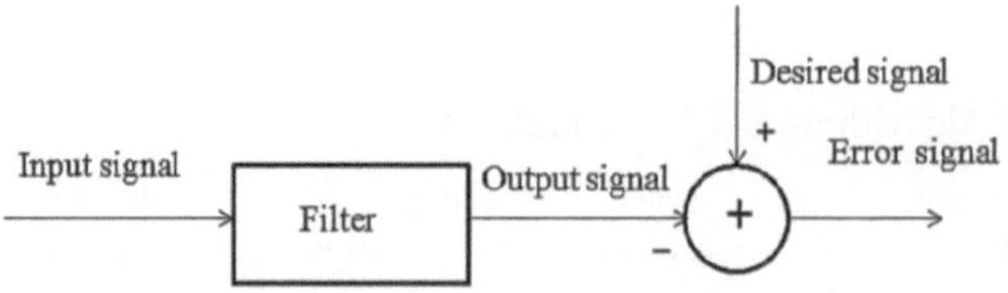

FIGURE 3.1 Basic schematic of a filter.

2. *Misadjustment:* This parameter provides a quantitative assessment that the ultimate value of the MSE deviates from the minimum MSE. It is calculated via a number of adaptive filters. This is formally defined as

$$M_{\text{ad}} = \frac{\text{MMSE}_{\text{SS}}}{\text{MMSE}} \tag{3.2}$$

where EMSE_{ss} is the steady-state value of excess mean square error (EMSE) and MMSE is the minimum mean square error. Misadjustment is a dimensionless factor that gives a metric of how near the algorithm is to optimal MSE.

The MMSE is the optimum filter error, which is a component of the primary signal not to cancel the ideal weight. After the algorithm reaches its stable state, it can be defined with alternating power of signal over samples. This can be expressed as

$$\text{MMSE} = \frac{1}{k-p}\sum_{n=P}^{k}|x(n)|^2 \tag{3.3}$$

Here, k is the overall sample amount and p is the amount of samples for which the algorithm reaches a stable state. EMSE is described as the result of weight deviation from their optimal values given by

$$\text{EMSE}(n) = \frac{1}{N}\sum_{j=0}^{N-1}|e_1(n-j)| \tag{3.4}$$

where $e_1(n)$ is the residual error and N is the amount of samples used within estimation. The steady-state EMSE is definite by taking EMSE over certain duration after the convergence characteristics are obtained. Its expression is given as

$$\text{EMSE}_{ss} = \frac{1}{k-p}\sum_{n=P}^{K}\text{EMSE}(k) \tag{3.5}$$

3. *Tracking:* As soon as an adaptive algorithm starts to function within a non-stationary environment, it needs to monitor environmental statistical variations. Two contradictory characteristics affect its tracking efficiency:

(i) convergence rate and (ii) fluctuations in steady state as a result of algorithmic noise.

4. *Computational requirements:* The concerns here include the number of operations to be performed to complete the algorithm's iteration. Normally in signal-processing applications, the algorithm's complexity is measured based on the amount of multiply and accumulate (MAC) computations required to perform the operation. Multiplication is more complex when compared to addition. So, in complexity reduction techniques, the number of multiplications has to be minimized.

3.2.3 Need for Development of Adaptive Exon Predictors

The comprehensive study within the region of bioinformatics remains finding exon sections within a gene sequence. In a species, the vital chromosomes form a subset that is necessary for growth, preservation, and reproduction. Therefore, the detection of exons in fresh pathogenic agents is of pragmatic significance for the discovery of natural illnesses and of drug targets. A genomic structure includes both protein-coding and non-protein-coding areas. The genomics subsection that relies on locating the protein coding areas in a genomic sequence is called gene identification.

The primary molecular structure divides living organisms into two types known as prokaryotes and eukaryotes. Protein-coding segments are constant and lengthy in prokaryotes; examples of prokaryotes are bacteria and archaea. A combination of eukaryotic coding sections divided by long non-protein-coding sections are termed as genes. These sections of protein coding are also known as exons, while non-protein-coding sections are known as introns [78]. All living organism are classified as Eukaryotes and whereas archaea and bacteria as prokaryotic organisms. The coding segments are only 3% of the genome in human eukaryotes, and the rest 97% are non-coding areas.

Identifying protein-coding segments is a major task [79]. DNA encoding areas show a Three Base Periodicity (TBP)property in nearly all DNA sequences. This is illustrated by a strong peak at frequencies $f = 1/3$ in power spectral density (PSD) diagram. In literature, various methods are outlined for finding exon regions according to various signal-processing methods. However, in real time, the sequence length is very long, and the exon positions vary from sequence to sequence. Current techniques of signal processing remain less accurate to estimate DNA coding areas. Methods to handle very long samples within several iterations are detected, and they can alter coefficients of weight according to the statistical nature of input sequence.

3.3 STRUCTURE OF ADAPTIVE EXON PREDICTOR FOR DNA ANALYSIS

A typical section diagram of the proposed AEP is shown in Figure 3.2. The sequence of genome is transformed to a numeric format in the proposed AEP. In the field of genomic signal processing (GSP), this is a significant task, as signal processing can only take place with digital or discrete signals. Also, the Voss representation is

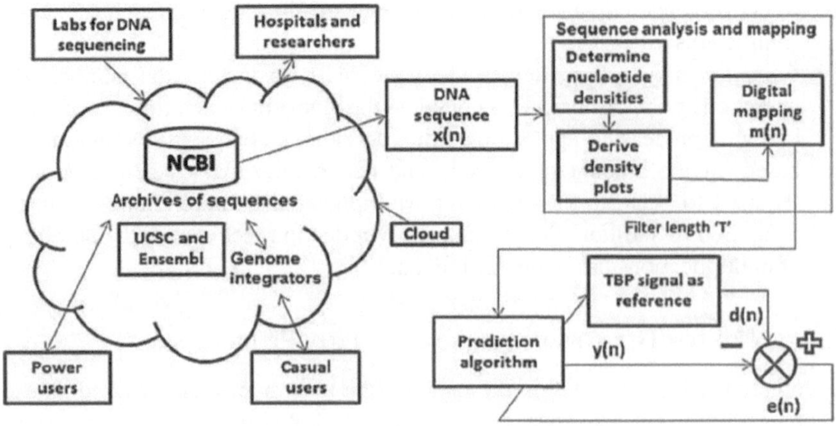

FIGURE 3.2 Structure of AEP for DNA analysis.

used at this point to convert a DNA sequence to a digital notation. The technique represents the input gene sequence as four binary sequences [80].

The occurrence of nucleotide is denoted as 1 and its absence as 0 in this binary mapping. The following set of numeric sequences can now be used as an input for the adaptive technique. Four numeric signals are used as the input to the adaptive filter. Now, we consider that the input for the AEP is transformed to four binary sequences. Let $x(n)$ be the DNA sequence, $M(n)$ a binary mapped sequence, $d(n)$ a TBP genomic sequence, $y(n)$ the output of adaptive algorithm, and $e(n)$ the signal for feedback for updating the weight coefficients of adaptive algorithm.

From Figure 3.2, it can be seen that an archive of sequences resulting from various sources such as gene sequencing labs, hospitals and researchers, power users and casual users are maintained in different gene databases such as NCBI, UCSC, and Ensembl in the cloud. Since most of the gene sequences are larger in length and size, we propose this cloud-based system to be used to acquire the input from the cloud in such cases, which saves a lot of time for further processing. Here, we have developed a novel AEP for DNA analysis using various adaptive signal-processing techniques as shown in the next part of the AEP structure.

In the AEP presented in the current work, the input DNA sequence from the NCBI databank is analyzed by determining nucleotide dimer densities with the help of density plots. The analyzed sequence is converted to a binary notation using a digital mapping technique. The numeric mapped sequence is provided as the input of prediction algorithms to locate the exon locations from the DNA sequences.

3.4 LMS ALGORITHM

The LMS algorithm is a two-process linear adaptive filtering algorithm.

1. *Filtering process:* The output of a filter generated by a certain number of inputs is computed, and an estimation error is generated by comparing this output to a desired response.

2. *Adaptive process:* The tap weights of the filter are automatically adjusted to reflect the estimated error.

A transversal filter design is generally used in adaptive filters as illustrated in Figure 3.3. Let $x(n)$ be assigned as the input to the filter n^2. The tap inputs $x(n)$, $x(n-1), x(n-2), \ldots x(n-T+1)$ form the elements of the T-by-1 tap input vector $x(n)$, where $L-1$ is the amount of delayed elements.

The tap weights $w(n) = \left[w_0, w_1, \ldots, w_{T-1} \right]^T$ thus form the elements of the T-by-1 tap-weight vector w with filter length T. The value calculated by LMS method for this vector is an estimate whose expected value is near to the wiener solution w_0 when the amount of iterations n reaches an infinite value. The mathematical modeling for LMS algorithm is presented in Table 3.2.

Instantaneous estimates for R and p are given by

$$R(n) = x(n)\ x^T(n)$$

$$p(n) = x(n)\ d^*(n)$$

Substituting estimates of R and p in $J(n)$, estimate of gradient vector $\nabla J(n)$ is given by

$$\nabla J(n) = -2p + 2R\ w(n) = -2\ x(n)d^*(n) + 2\ x(n)x^T(n)w(n)$$

Also, the expression for $J(n)$ is given as

$$J(n) = E\left\{ |e(n)|^2 \right\} = E\left\{ |d(n) - y(n)|^2 \right\} \tag{3.8}$$

where the error $e(n)$ is defined as the difference between the desired response $d(n)$ and the actual response $y(n)$.

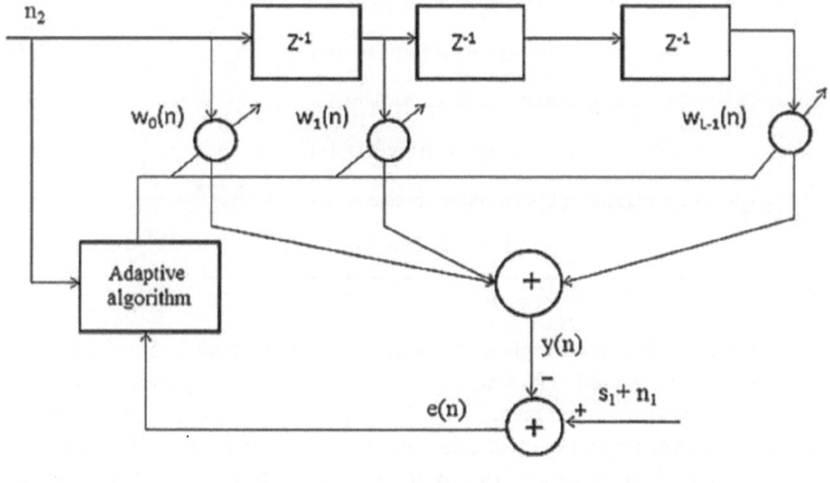

FIGURE 3.3 Structure of generalized transversal filter.

TABLE 3.2
Mathematical Modeling of LMS Algorithm

Parameters: T = number of taps (i.e. filter length)

μ = step size parameter

Let the tap input be $x(n)$ and filter length T is moderate to large.

Initialization: Set $w(0) = 0$ as the initial condition.

Data: Given $x(n)$ = T-by-1 tap input vector to filter $n2$ at time n

$$= \left[x(n), x(n-1) \ldots x(n-T+1)\right]^T$$

$d(n)$ = desired response at time n, R = correlation matrix of $x(n)$, p = cross-correlation vector between $x(n)$ and $d(n)$, σ_d^2 is the variance of desired response $d(n)$

To be computed: $w(n+1)$ = estimate of tap-weight vector at time $n+1$

Computation:

The FIR filter output is the inner product of $x(n)$ and $w(n)$ is given by

$$y(n) = x^T(n)\, w(n) = w^T(n)\, x(n) \tag{3.6}$$

The expression for estimation error $e(n)$ is given by

$$e(n) = d(n) - y(n) = d(n) - w^T(n).x(n) \tag{3.7}$$

The weight vector $w(n+1)$ corresponds to the current weight vector $w(n)$ plus a shift proportional to negative gradient is expressed as

$$w(n+1) = w(n) - \mu \nabla J(n) \tag{3.9}$$

The new recursive relation for $\nabla J(n)$ is written as

$$\nabla J(n) = \nabla E\left\{\left|e(n)\right|^2\right\} = E\left\{\nabla\left|e(n)\right|^2\right\} = E\left\{e(n)\nabla e^*(n)\right\} \tag{3.10}$$

$$\nabla e^*(n) = -x^*(n) \tag{3.11}$$

Substituting (3.11) in (3.10), resultant expression is given by

$$\nabla J(n) = -E\left\{e(n)x^*(n)\right\} \tag{3.12}$$

Now substituting (3.12) in (3.9), the weight recursion expression is given by

$$w(n+1) = w(n) + \mu E\left\{e(n)x^*(n)\right\} \tag{3.13}$$

In the case of LMS, 1-point sample mean with $L = 1$, such that

$$\hat{E}\left\{e(n)x^*(n)\right\} = e(n)x^*(n) \tag{3.16}$$

Thus, the weight update equation of LMS algorithm becomes

$$w(n+1) = w(n) + \mu\, e(n)\, x(n) \tag{3.17}$$

The adaptive LMS algorithm is a practical way to find a near approximate in real time. Accuracy is restricted to its statistical sample size. The steepest descent method was used by LMS for its implementation. The step size μ influences the rate at which weight vector moves down the quadratic area. The adjustment applied to $w(n)$ is low for a small value of step size μ. Also, by increasing the μ, the rate of descent rises.

However, gradient noise results from the recursive calculation of each tap weight of LMS algorithm.

The unknown expectation $E\left[e(n)x^*(n)\right]$ replaced with an estimate such as sample mean can be expressed as

$$\hat{E}\{e(n)x^*(n)\} = \frac{1}{L}\sum_{l=0}^{L-1} e(n-l)x^*(n-l) \tag{3.14}$$

Incorporating this estimation into the steepest decent, the weight update equation becomes

$$w(n+1) = w(n) + \mu\frac{1}{L}\sum_{l=0}^{L-1} e(n-l)x^*(n-l) \tag{3.15}$$

The LMS flow chart is as shown in Figure 3.4. This algorithm converges in mean and remains robust until the step size parameter μ of the variable autocorrelation matrix R_x exceeds 0 but is less than two times the reciprocal of its biggest eigen value $\lambda_{\max}\left(0 < \mu < \frac{2}{\lambda_{\max}}\right)$.

The problems with LMS algorithm are selecting the step size parameter μ, slow convergence, and small steady-state error. In contrast, convergence is quick for a larger value of μ, but the steady-state error is greater. So, it is hard in LMS algorithms to select the correct value of μ. The step-size parameter μ depends on the input signal power, and it is more when the weight vector converges in mean; coefficients start to fluctuate with optimum values.

These changes are caused by noisy gradient vectors used to make $w(n)$ adjustments. This means that the weight error vector variance is not zero and the MSE is larger by an amount called the excess MSE than the minimum MSE. This has demonstrated a continuous weight vector bias for convergence evaluation of LMS algorithm with a deterministic reference input.

3.5 LMF ALGORITHM

Adaptive filters are more efficient than the mean square estimate in specific situations used for LMS algorithm relying on higher-order statistics. To investigate this, we have created various AEPs to locate exon locations in gene sequences. The least mean fourth (LMF) algorithm remains one such instance because of the minimization of the fourth moment of output estimate error [81]. This algorithm remains closely related to LMS algorithm. LMF algorithm tries to a higher power of the error signal, specifically the fourth order. The mathematical modeling for LMF algorithm is presented in Table 3.3.

The Weiner solution for a linear estimation problem is given by

$$d(n) = w_0^T\, X(n) + e_0(n)$$

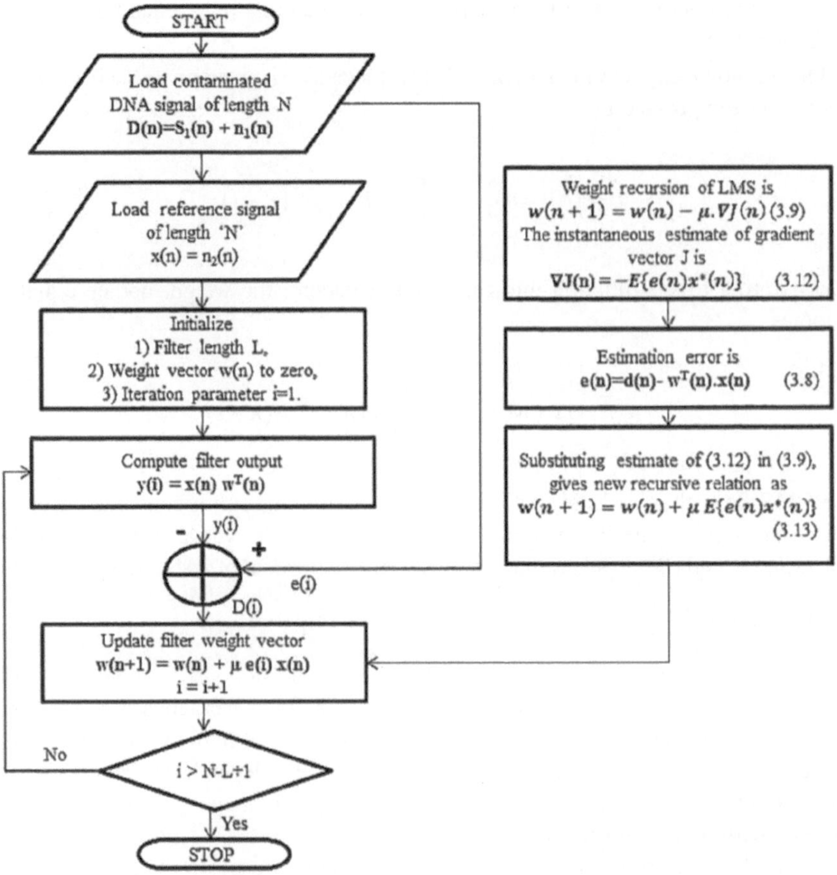

FIGURE 3.4 Flow chart for LMS algorithm.

From the steepest descent algorithm, the expression for gradient vector is given by

$$\nabla J(n) = -2p + 2R\ w(n) = -2\ x(n)d^*(n) + 2\ x(n)x^T(n)w(n)$$

The expression for $w(n + 1)$ from the steepest descent algorithm is written as

$$w(n+1) = w(n) - \mu \nabla J(n)$$

The unknown expectation $E\left[e^3(n)x^*(n)\right]$ is replaced with an estimate such as sample mean given by

$$\hat{E}\left\{e(n)x^*(n)\right\} = \frac{1}{L}\sum_{l=0}^{L-1} e^3(n-l)x^*(n-l) \tag{3.21}$$

TABLE 3.3
Mathematical Modeling of LMF Algorithm

Parameters: T = number of taps (i.e. filter length)

μ = step size parameter

Let the tap input be $x(n)$ and filter length T is moderate to large.

Initialization: Set $w(0) = 0$ as the initial condition.

Data: Given $x(n)$ = T-by-1 tap input vector to filter n_2 at time n

$$= [x(n), x(n-1) \dots x(n-T+1)]^T$$

$X(n)$ is unity variance vector, $d(n)$ = desired response at time n, R = correlation matrix of $x(n)$, p = cross-correlation vector between $x(n)$ and $d(n)$, σ_d^2 is the variance of desired response $d(n)$, and $(.)^T$ is the transpose of (.).

To be computed:

$w(n+1)$ = estimate of tap-weight vector at time $n+1$

Computation:

Output of FIR filter is given by

$$y(n) = x^T(n)\, w(n) = w^T(n)\, x(n)$$

The expression for estimation error e(n) is given by

$$e(n) = d(n) - y(n) = d(n) - w^T(n).x(n)$$

The expression for cost function of LMF is $J(n) = E\left[|e(n)|\right]^4$.

The new recursive relation for $\nabla J(n)$ is written as

$$\nabla J(n) = \nabla E\left[|e(n)|\right]^4 = E\left\{\nabla\left[|e(n)|\right]^4\right\} = E\left\{e^3(n)\nabla e^*(n)\right\} \tag{3.18}$$

Substituting (3.11) in (3.18), resultant expression is given by

$$\nabla J(n) = -E\left\{e^3(n)\, x^*(n)\right\} \tag{3.19}$$

Now substituting (3.19) in (3.9), the weight recursion expression is given by

$$w(n+1) = w(n) + \mu\ E\left\{e^3(n)x^*(n)\right\} \tag{3.20}$$

In the case of LMS, 1-point sample mean with $L = 1$, such that

$$\hat{E}\left\{e^3(n)\, x^*(n)\right\} = e^3(n)\, x^*(n) \tag{3.23}$$

Therefore, the weight update equation of LMF algorithm becomes

$$w(n+1) = w(n) + \mu e^3(n)x(n) \tag{3.24}$$

Incorporating this estimation into the steepest decent, weight update equation becomes

$$w(n+1) = w(n) + \mu \frac{1}{L} \sum_{l=0}^{L-1} e^3(n-l)x^*(n-l) \tag{3.22}$$

The LMF algorithm gives less error in steady state than the standard LMS algorithm. This is because of the dependency of EMSE of LMS on the second-order moment of noise. In comparison, the surplus EMSE of the LMF algorithm relies on higher-order moments of noise which leads to a less steady-state error than LMS algorithm. The flow chart of the LMF algorithm is as shown in Figure 3.5. It was also known that this feature of the LMF algorithm is recognized by the fact that the convergence behavior of LMS and LMF algorithms is susceptible to the condition number, i.e. in relation to the eigen ratio of the autocorrelation input signal matrix.

The data normalization can be used when the LMF recursion has been altered to include an inverse norm of the output regressor, to solve this result of the proportion.

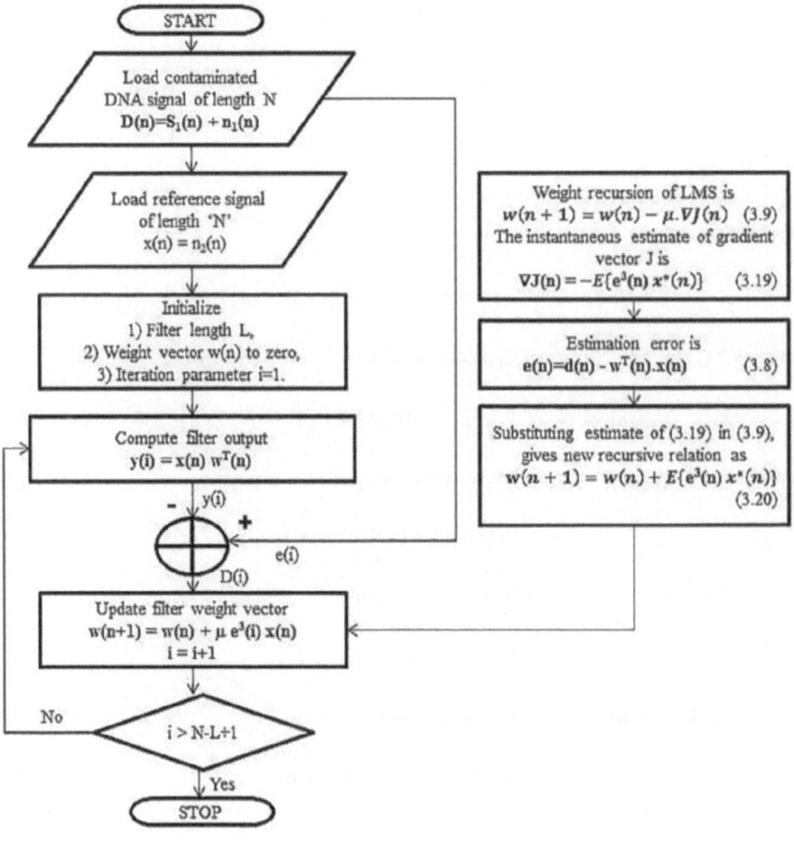

FIGURE 3.5 Flow diagram for LMF algorithm.

This leads to a good stability and offers a quicker convergence. However, in order to achieve a stable adaptation, this high-order statistics of error in the LMF algorithm requires lower step size.

3.6 VARIABLE STEP SIZE LMS (VSLMS) ALGORITHM

An easy way to increase the efficiency of LMS algorithm is to have a varying step size rather than fixed during its initial convergence. Also, by using small step size values near its steady state, this results in variable step size LMS (VSLMS) algorithm. The rapid convergence rate and a small steady-state MSE can be achieved by using such a strategy.

When the instantaneous error is positive or negative, the step size is increased or decreased by a small value in VSLMS algorithms. This algorithm utilizes a variable move to decrease the trade-off among tracking ability and misadjustment of fixed step size LMS algorithm. Thus, VSLMS is a robust, simple, and efficient algorithm characterized by a fast convergence and a low steady-state mean squared error [82,83]. The flow diagram of VSLMS algorithm is clearly depicted in Figure 3.6. The mathematical modeling for VSLMS algorithm is presented in Table 3.4.

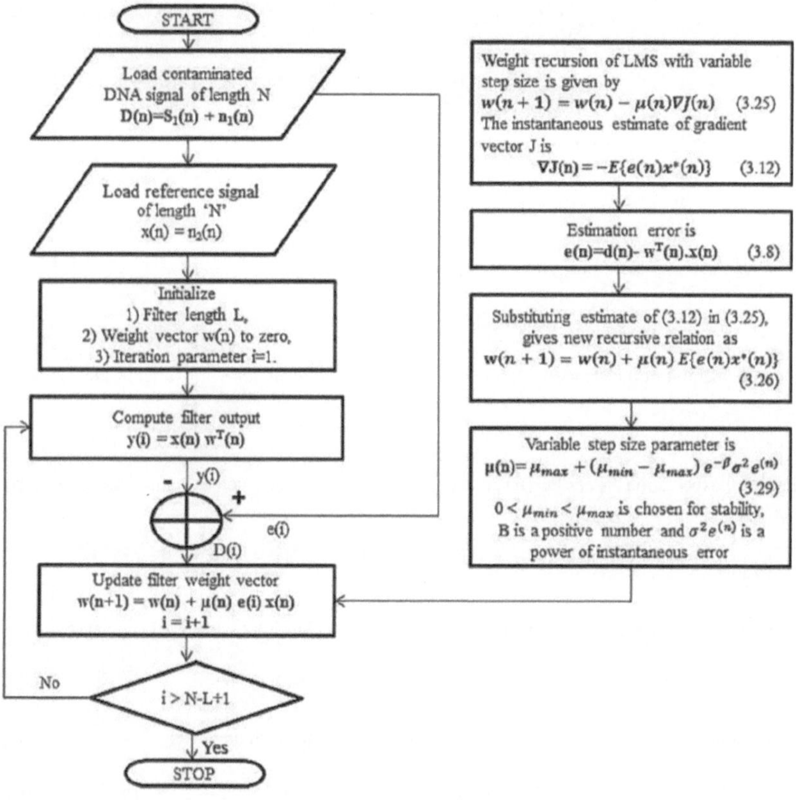

FIGURE 3.6 Flow chart for VSLMS algorithm.

TABLE 3.4
Mathematical Modeling of VSLMS Algorithm

Parameters: T = number of taps (i.e. filter length), μ = step size parameter
Let the tap input be $x(n)$ and filter length T is moderate to large.
Initialization: Set $w(0) = 0$ as the initial condition.
Data: Given $x(n)$ = T-by-1 tap input vector to filter n_2 at time n

$$= \left[x(n), x(n-1) \ldots x(n-T+1)\right]^T$$

$d(n)$ = desired response at time n, $\sigma^2 e^{(n)}$ is the power of instantaneous error, β is a positive integer, σ^2 is the variance of desired response $d(n)$, and $(.)^T$ is the transpose of (.).
To be computed:
$w(n + 1)$ = estimate of tap-weight vector at time $n + 1$
Computation:
The Output of FIR filter is given by

$$y(n) = x^T(n)w(n) = w^T(n)x(n)$$

The expression for estimation error $e(n)$ is given by

$$e(n) = d(n) - y(n) = d(n) - w^T(n) \cdot x(n)$$

The new recursive relation for $\nabla J(n)$ is written as

$$\nabla J(n) = E\left\{\nabla \left|e(n)\right|^2\right\} = \nabla E\left\{\left|e(n)\right|^2\right\} = E\left\{e(n)\nabla e^*(n)\right\}$$

and

$$\nabla e^*(n) = -x^*(n)$$

Thus, the resultant expression for gradient vector is given by

$$\nabla J(n) = -E\left\{e(n)x^*(n)\right\}$$

Based on the method of the steepest descent, the weight update recursion with variable step size $\mu(n)$ is

$$w(n+1) = w(n) - \mu(n)\nabla J(n) \tag{3.25}$$

The unknown expectation $E\left[e(n)x^*(n)\right]$ is replaced with an estimate such as sample mean given by

$$\hat{E}\left\{e(n)x^*(n)\right\} = \frac{1}{L}\sum_{l=0}^{L-1} e(n-l)x^*(n-l) \tag{3.26}$$

Therefore, from (3.12), the weight update relation of VSLMS algorithm with $\mu(n)$ is given by

$$w(n+1) = w(n) + \mu(n)\, x(n)e(n) \tag{3.28}$$

Thus, the variable step size parameter $\mu(n)$ of VSLMS algorithm is calculated as

$$\mu(n) = \mu_{\max} + (\mu_{\min} - \mu_{\max})e^{-\beta}\sigma^2 e^{(n)} \tag{3.29}$$

$w(n)$ is the tap weight vector with the taps stored in a row vector

$$= \left[w(n) \ w(n-1) \ldots \ w(n-T+1) \right]^T$$

Incorporating the above estimation into the steepest decent, the resulting weight update equation becomes

$$w(n+1) = w(n) + \mu(n)\frac{1}{L}\sum_{l=0}^{L-1} e(n-l)x^*(n-l) \tag{3.27}$$

where $\mu(n)$ is the variable step size calculated with an appropriate formula.

Thus, the weight update equation of LMS algorithm becomes

$$w(n+1) = w(n) + \mu \ e(n) \ x(n)$$

In (3.29), from the flow chart of VSLMS, $0 < \mu_{\min} < \mu_{\max}$ is selected to offer minimum tracking capability and to guarantee algorithm stability. The parameter β is a positive number that adds some flexibility of algorithm design and $\sigma^2 e^{(n)}$ is a power of instantaneous error. These changes are caused by the noisy gradient vectors used to correct $w(n)$.

3.7 LEAST MEAN LOGARITHMIC SQUARES (LMLS) ALGORITHM

To improve the convergence and stability performance of AEP, we have proposed the use of adaptive techniques depending upon comparative logarithmic costs termed as LMLS algorithm. This technique overcomes LMS setbacks and increases the speed of convergence and tracking performance.

$w(n)$ is the tap weight vector with the taps stored in a row vector

$$= \left[w(n) \ w(n-1) \ldots \ w(n-T+1) \right]^T$$

Here, an unknown vector ω_0 is presented with a linear model as

$$d(n) = \omega_0^T x(n) + n_t \tag{3.30}$$

where the input signal is denoted by $x(n)$ and n_t represents the noise.

The error signal is defined as $e(n)$ where $w(n)$ is the weight vector of adaptive filter (Table 3.5).

The adaptive filtering algorithms here assess the unknown system vector by minimizing the cost function. The techniques for gradient descent generally use convex and uni-modal cost functions to converge close to a global minimum of error surfaces. The different powers of $e(n)$ presented and a linear combination of different error powers are also widely used. The optimal error non-linear function can be expressed as a linear combination of different orders of error, and for the optimum error, non-linearity of the algorithms can be more approximate to any individual

TABLE 3.5
Mathematical Modeling of LMLS Algorithm

Parameters: T = number of taps (i.e. filter length),

μ = step size parameter, Let the tap input be $x(n)$ and filter length T is moderate to large.

Initialization: Set $w(0) = 0$ as the initial condition.

Data: Given $x(n) = T$-by-1 tap input vector to filter n_2 at time n

$$= \left[x(n), x(n-1)\ldots x(n-T+1)\right]^T$$

where $w(n)$ is the tap weight vector, $d(n)$ = desired response at time n, ω_0 is an unknown vector, and $(.)^T$ is the transpose of (.).

To be computed:

$w(n + 1)$ = estimate of tap-weight vector at time $n + 1$

Computation:

The FIR filter output is given by $y(n) = x^T(n)w(n) = w^T(n)x(n)$

The conventional cost function of the error signal $e(n)$ is $F\left[e(n)\right] = E\left[\left(e(n)\right)^2\right] = E\left[\left|e(n)\right|\right]$.

Here, the normalized error cost function introduced using logarithmic function is given by

$$J\left(e(n)\right) = F\left(e(n)\right) - \frac{1}{\alpha}\, ln(1 + \alpha F\left(e(n)\right) \tag{3.31}$$

where $\alpha > 0$ is a design parameter and $F\left(e(n)\right)$ is a conventional cost function of error signal $e(n)$. Thus, the resultant expression for gradient vector is given by

$$\nabla J(n) = -E\left\{e(n)x^*(n)\right\}$$

Based on the method of the steepest descent, the weight update recursion with variable step size $\mu(n)$ is

$$w(n+1) = w(n) - \mu(n)\nabla J(n) \tag{3.25}$$

Based on the gradient of $J(e(n))$, the general steepest descent update is obtained as

$$w(n+1) = w(n) - \Delta_w \cdot F\left(e(n)\right)\left[\frac{\alpha F\left(e(n)\right)}{1 + \alpha F\left(e(n)\right)}\right] \tag{3.32}$$

where $\Delta_w \cdot F\left(e(n)\right)$ is the first gradient of (3.11), $\mu > 0$ is the step size, and $\alpha > 0$ is the design parameter. Then, the stochastic gradient update is given by

$$w(n+1) = w(n) + \mu \cdot x(n)\frac{\partial f\left(e(n)\right)}{\partial e(n)}\left[\frac{\alpha f\left(e(n)\right)}{1 + \alpha f\left(e(n)\right)}\right] \tag{3.33}$$

Thus, for $F\left(e(n)\right) = E\left(e(n)\right)^2$, the weight update equation of the LMLS algorithm becomes

$$w(n+1) = w(n) + \mu \cdot x(n)e(n)\left[\frac{\alpha\left(e(n)\right)^2}{1 + \alpha\left(e(n)\right)^2}\right] \tag{3.34}$$

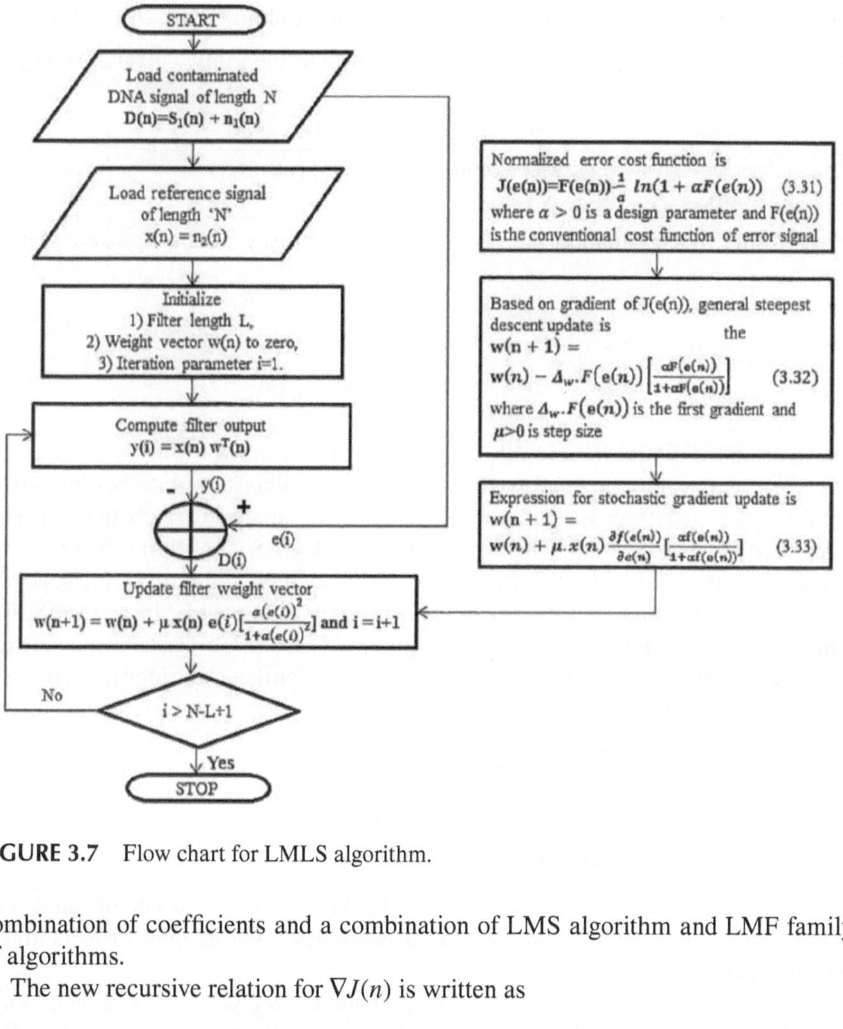

FIGURE 3.7 Flow chart for LMLS algorithm.

combination of coefficients and a combination of LMS algorithm and LMF family of algorithms.

The new recursive relation for $\nabla J(n)$ is written as

$$\nabla J(n) = E\left\{\nabla |e(n)|^2\right\} = \nabla E\left\{|e(n)|^2\right\} = E\left\{e(n)\nabla e^*(n)\right\}$$

$$\nabla e^*(n) = -x^*(n)$$

Thus, the resultant expression for gradient vector is given by

$$\nabla J(n) = -E\left\{e(n)x^*(n)\right\}$$

Also, the first gradient of the relation in (3.11) is given by

$$\Delta_w \cdot F\left(e(n)\right)$$

Here, the expectation $E\left[e(n)\,x^*(n)\right]$ is generally unknown in this algorithm. Therefore, it must be replaced with an estimate such as the sample mean given by

$$E\left\{e(n)x^*(n)\right\} = \frac{1}{T}\sum_{l=0}^{T-1}\frac{\partial f(e(n-l))}{\partial e(n-l)}\left[\frac{\alpha f(e(n-l))}{1+\alpha f(e(n-l))}\right]x^*(n-l)$$

Incorporating this estimation into the steepest decent, the weight update equation becomes

$$w(n+1) = w(n) + \mu\frac{1}{T}\sum_{l=0}^{T-1}\frac{\partial f(e(n-l))}{\partial e(n-l)}\left[\frac{\alpha f(e(n-l))}{1+\alpha f(e(n-l))}\right]x^*(n-l)$$

The algorithm is similar to a lowest-average fourth revision with tiny mistake numbers while acting like the lowest-average rectangular algorithm with big input disturbances. The fourth-order statistics of the error for smaller perturbations gives a lower steady-state MSE [84]. The flow diagram for LMLS algorithm is as shown in Figure 3.7. Thus, LMLS algorithm combines LMS and LMF algorithms intrinsically based on the error amount, rather than the mixed algorithms of LMF and LMS that require an artificial parameter in cost definition. LMLS achieves the same convergence performance with LMF algorithm and also significantly improves stability [86]. This algorithm also attains similar trade-off between the transient and steady-state performances of LMF algorithm.

3.8 LEAST LOGARITHMIC ABSOLUTE DIFFERENCE (LLAD) ALGORITHM

The least logarithmic absolute difference (LLAD) technique gradually and elegantly adapts conventional cost function depending on the amount of error in its implementation [85].

In impulse-free noise environments, LLAD and LMS algorithms exhibit a likely convergence behavior, while LLAD algorithm is robust against impulsive interference and exceeds sign algorithm. The flow diagram for LLAD algorithm is as shown in Figure 3.8.

Parameters: T = number of taps (i.e. filter length)

μ = step size parameter

Let the tap input be $x(n)$ and filter length T is moderate to large.

Initialization: Set $w(0) = 0$ as initial condition.

Data: Given $x(n)$ = T-by-1 tap input vector to filter n^2 at time n

$$= \left[x(n), x(n-1) \;....\; x(n-T+1)\right]^T$$

$w(n)$ is the tap weight vector, $d(n)$ is the desired response at time n, ω_0 is an unknown vector, and $(.)^T$ is the transpose of (.).

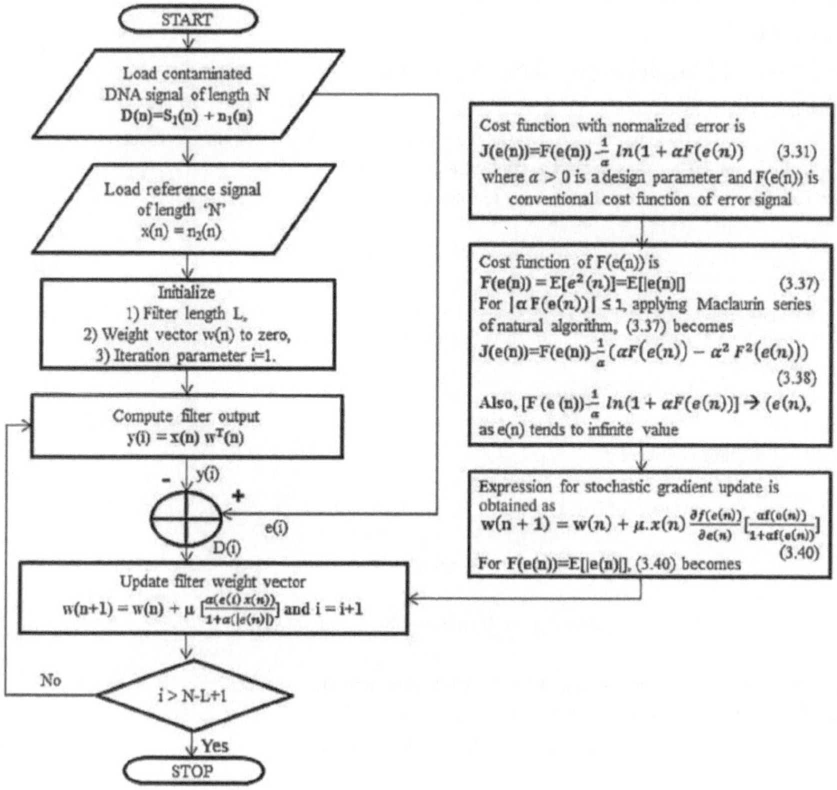

FIGURE 3.8 Flow diagram for LLAD algorithm.

To be computed:
$w(n + 1)$ = estimate of tap-weight vector at time $n + 1$
Computation:
Here, an unknown vector ω_0 is presented with a linear model as

$$d(n) = \omega_o^T x(n) + n_t$$

The instantaneous estimate of gradient vector J is written as

$$\nabla' J(n) = -\, 2\ x(n)\ d^*(n) + 2\ d(n)\ x^T(n)w(n)$$

where $x(n)$ is the tap input vector and $w(n)$ is the tap weight vector (Table 3.6).

Here, tap weight vector $w(n)$ is a random vector depends on tap-input vector $x(n)$ with its taps stored in a row vector given by

$$= \left[w(n)\ w(n-1)\ldots\ w(n-T+1)\right]^T$$

TABLE 3.6
Mathematical Modeling of LLAD Algorithm

Parameters: T = number of taps (i.e. filter length),

μ = step size parameter, Let the tap input be $x(n)$ and filter length T is moderate to large.

Initialization: Set $w(0) = 0$ as the initial condition.

Data: Given $x(n)$ = T-by-1 tap input vector to filter n_2 at time n

$$= \left[x(n), x(n-1) \ldots x(n-T+1)\right]^T$$

$w(n)$ is the tap weight vector, $d(n)$ is the desired response at time n, ω_0 is an unknown vector, σ_d^2 is the variance of desired response $d(n)$, and $(.)^T$ is the transpose of (.).

To be computed:

$w(n + 1)$ = estimate of tap-weight vector at time $n + 1$

Computation:

Output of FIR filter is given by

$$y(n) = \left[w_0(n)x(n) + \ldots + w_{L-1}(n)x(n-M+1)\right] = w^T(n)x(n)$$

Here, the normalized error cost function introduced using logarithmic function is given by

$$J(e(n)) = F(e(n)) - \frac{1}{\alpha}\ln\left(1 + \alpha F(e(n))\right)$$

where $\alpha > 0$ is a design parameter and $F(e(n))$ is a conventional cost function of error signal $e(n)$. Thus, the resultant expression for gradient vector is given by $\nabla J(n) = -E\{e(n)x^*(n)\}$.

Based on the method of the steepest descent, the weight update recursion with variable step size $\mu(n)$ is

$$w(n+1) = w(n) - \mu(n)\nabla J(n)$$

For $\left|\alpha F(e(n))\right| \leq 1$, applying Maclaurin series using natural algorithm, Equation (3.37) gives

$$J(e(n)) = F(e(n)) - \frac{1}{\alpha}\left(\alpha F(e(n)) - \alpha^2 F^2(e(n))\right) \tag{3.38}$$

For smaller values of error, cost function $J(e(n))$ resembles $F(e(n))$, for instance,

$$\left[F(e(n)) - \frac{1}{\alpha}\ln\left(1 + \alpha F(e(n))\right)\right] F(e(n)) \tag{3.39}$$

as $e(n)$ tends to an infinite value.

Thus, the general update expression of stochastic gradient is stated as

$$w(n+1) = w(n) + \mu.x(n).\frac{\partial f(e(n))}{\partial e(n)}\left[\frac{\alpha f(e(n))}{1 + \alpha f(e(n))}\right] \tag{3.40}$$

Therefore, the relation for mass update expression for LLAD algorithm is

$$w(n+1) = w(n) + \mu \cdot \left[\frac{\alpha(e(n)x(n))}{1 + \alpha(|e(n)|)}\right] \tag{3.42}$$

The expression for estimation error is given by

$$e(n) = d(n) - w^T(n)\, x(n)$$

where the term $w^T(n)\, x(n)$ is the inner product of $w(n)$ and $x(n)$.

The new recursive relation for $\nabla J(n)$ is written as

$$\nabla J(n) = E\left\{\nabla |e(n)|^2\right\} = \nabla E\left\{|e(n)|^2\right\} = E\left\{e(n)\nabla e^*(n)\right\}$$

$$\nabla e^*(n) = -x^*(n)$$

Also, the first gradient of the relation in (3.11) is given by $\Delta_w \cdot F(e(n))$.

The signum representation is given below:

$$\text{Sign}\{x(n)\} = \left\{ \begin{array}{c} 1 : x(n) > 0 \\ 0 : x(n) = 0 \\ -1 : x(n) < 0 \end{array} \right\} \tag{3.35}$$

These signed variants are preferable to reduce the computational difficulty of LMS. Among all signed variants, the sign regressor version offers a low computational complexity with a less number of multiplications.

Here, sign regressor LLAD (SRLLAD) technique remains obtained as a recursion of LMS to alter tap input vector by applying sign function on each element. To further reduce complexity in computations, normalized LLAD algorithm is derived in Chapter 5.

The step size limit for the convergence of LMS algorithm is determined by

$$0 < \mu < \frac{2}{x^T(n)x(n)} \tag{3.36}$$

Substituting the estimate of $\nabla' J(n)$ in the steepest descent algorithm, new recursive relation for updating tap weight vector is

$$w(n+1) = w(n) + \mu x(n)\left[d^*(n) - x^T(n).w(n)\right]$$

LLAD overwhelms shortcomings and increases convergence speed along with exon identification capability compared to LMS. LLAD-based AEP is presented to prevent the weight drift problem of LMS. The cost function $F(e(n)) = E[|e(n)|]$ is utilized by sign algorithm (SA), which provides robustness against impulsive interferences.

Thus, the cost function of $F(e(n))$ can be written as

$$F(e(n)) = E\left[e^2(n)\right] = E\left[|e(n)|\right] \tag{3.37}$$

For lesser values of $F(e(n))$, it is an infinite combination for conventional cost functions. Therefore, the logarithmic techniques use amalgamation of cost functions $F(e(n))$ and $F^2(e(n))$ depending on the amount of error.

For $F(e(n)) = E[|e(n)|]$ in (3.40), the resultant expression becomes

$$w(n+1) = w(n) + \mu.x(n).\ \text{sign}(e(n))\left[\frac{\alpha(|e(n)|)}{1+\alpha(|e(n)|)}\right] \tag{3.41}$$

As norm is the power of least probable error aimed at convex cost function, SA delivers a slow rate of convergence. The SA, however, has a faster rate of convergence since the norm for convex cost functions has the lowest possible error.

3.9 SIMPLIFIED ALGORITHMS BASED ON SIGNUM FUNCTION

Studying the composition of main protein regions enables the secondary and tertiary structures of exon segments to detect whole abnormalities, heal illnesses, and develop drugs once full composition of protein areas is evaluated [86,87]. These surveys promote knowledge assessment of phylogenic trees [88]. Novel algorithms that employ the sign function to either error, input signal, or both were derived for simplicity of implementation from different algorithms depending on the LMS discussed in the above sections. This significantly reduces the calculation time, in particular, for MAC computations. Use of signum function reduces computational complexity for adaptive filter and therefore is appropriate for GSP applications. In this section, we apply signum function to various algorithms and evaluate the performance in locating the exon locations from the DNA sequence.

The sign-based algorithms enjoy the advantages of their LMS-based versions, mentioned in the previous sections and reduced computational complexity due to signum function present in the algorithm. Adaptive filter efficiency relies primarily on the algorithm used for weight updating. Convergence analysis for LMS and signed algorithms leads to a conclusion that LMS algorithms are exponentially convergent, whereas the signed algorithms are linearly convergent. This makes the convergence of signed algorithms slowest when the initial weight setting is far from the optimum. Various existing gene predictors, methods based on discrete wavelet transform and cross-correlation, and several computational methods are discussed in Refs. [89–91].

3.9.1 Sign-Based LMS Algorithms

The derivation of these algorithms is similar to LMS technique up to a certain level. At this stage, recursion remains modified thru applying signum function. SRA, SA, and SSA are most significant in this category.

i. Sign Regressor LMS (SRLMS) algorithm

   SRLMS remains achieved by combining a $x(n)$ tap input matrix with a Sign$\{x(n)\}$ matrix, where a sign feature has been introduced to vector $x(n)$ on a component-by-component basis. This is also called as clipped LMS since the input data is clipped. The weight recursion expression for SRA is given by

$$w(n+1) = w(n) + \mu\, e(n)\text{Sign}\{x(n)\} \tag{3.43}$$

where $\text{Sign}\{x(n)\} = \left\{ \begin{array}{c} 1 : x(n) > 0 \\ 0 : x(n) = 0 \\ -1 : x(n) < 0 \end{array} \right\}$

The kth coefficient in signed data vector could be expressed as

$$\text{Sign}\{x(n-k)\} = \frac{x(n-k)}{|x(n-k)|} \tag{3.44}$$

In normalized LMS technique, wherever normalization of weight vector is done with the same weight vector $\|x(n)\|^2$, the SRA is individually normalized for each weight vector coefficient which works better than other LMS algorithms.

ii. Sign LMS (SLMS) algorithm

The variant is acquired from the standard recursion of LMS thru combining the $e(n)$ by sign function [92], which is also referred to as the pilot LMS or sign LMS algorithm. This results in weight recursion equation given by

$$w(n+1) = w(n) + \mu\text{Sign}\{e(n)\}x(n) \tag{3.45}$$

Due to the substitution of $e(n)$, recursion is less costly than the standard LMS recursion, in particular in high-speed applications where hardware implementation may be necessary. This simplification in the sign algorithm (SA) occurs when the selected step size is power of two, i.e. $\mu = 2 - 1$, so that no multiply operation is necessary to implement recursion. Thus, the filter tap weight would be updated by a number of shifts and add/subtract operations.

iii. Sign-Sign LMS (SSLMS) algorithm

This SSLMS technique combines recursions of SA and SRA which results in the below weight recursion expression:

$$w(n+1) = w(n) + \mu\text{Sign}\{e(n)\}\text{Sign}\{x(n)\} \tag{3.46}$$

It is otherwise termed as zero forcing LMS as it needs zero multiply computations to implement [93]. Its output is mildly worse than the standard regressor method. The signing and signing schemes are, however, both much lighter than the standard LMS matrix. They also have a very strange convergence conduct. At first, they start very smoothly, but then accelerate as the amount of MSE falls. This is illustrated below.

The weight recursion relation for SA is given by

$$w(n+1) = w(n) + \mu\{x(n)\}\{e(n)/|e(n)|\} \quad (3.47)$$

where Sign[$e(n)$] = $e(n)/|e(n)|$.

The relation is reorganized as follows:

$$w(n+1) = w(n) + \left[\frac{\mu}{|e(n)|}\right]x(n)e(n) \quad (3.48)$$

It is found in this equation that an SA using varying step parameter $\mu'(n) = \{\mu/|e(n)|\}$ may be considered as LMS algorithm. As the sign algorithm converges step parameter, $\mu'(n)$ rises with a decrease in the $e(n)$ value. A very low step size needs to be used to maintain the SA consistent using a minimum error at steady state. A tiny μ within the original part of the tag matrix contributes to a similarly tiny valuation of $\mu'(n)$. Hence, this slowly converges the algorithm at first.

However, with the algorithm converging and $e(n)$ becoming lesser, $\mu'(n)$ is higher and results in fast convergence. In addition, hardware implementation becomes simple due to the sign present in the algorithm. Likewise, it is possible to analyze the behavior of SRA. In this situation, a distinct variable step parameter is used to control every filter tap. For instance, the step size of ith filter tap at the nth iteration is

$$\mu_i'(n) = \{\mu/|x(n-i)|\} \quad (3.49)$$

where μ is the common parameter to all taps.

The changes in variable step parameter $\mu_i'(n)$ are not dependent on the convergence of filter. In this case, the step size selection is based on the average size of $x(n)$. The SRA thus results in a more uniform convergence compared to the SA.

3.10 EXTENSION TO SIGN-BASED REALIZATIONS OF LMS-BASED VARIANTS

3.10.1 Sign-Based Least Mean Fourth (LMF) Algorithms

LMF technique is a basic version of higher-order adaptive filter. Combining the LMF in (3.24) with SRA, SA, and SSA results in its signed variants SRLMF, SLMF, and SSLMF, respectively. The expressions of weight recursion of these techniques are given as follows:

$$w(n+1) = w(n) + \mu\, e^3(n)\text{Sign}\{x(n)\} \quad (3.50)$$

$$w(n+1) = w(n) + \mu\text{Sign}\{e^3(n)\}x(n) \quad (3.51)$$

$$w(n+1) = w(n) + \mu\text{Sign}\{e^3(n)\}\text{Sign}\{x(n)\} \quad (3.52)$$

From (3.51) and (3.52), it is noted that weight recursive expressions for SLMF and SSLMF are the same as SLMS and SSLMS algorithms.

3.10.2 Sign-Based Variable Step Size LMS (VSLMS) Algorithms

VSLMS technique is a basic variant of higher-order adaptive filter [94]. Combining the VSLMS in (3.28) with SRA, SA, and SSA algorithms results in VSRLMS, VSLMS, and VSSLMS algorithms, respectively. The weight relations of VSLMS-based signed variants are written as follows:

$$w(n+1) = w(n) + \mu(n)\ \text{Sign}\{x(n)\}e(n) \tag{3.53}$$

$$w(n+1) = w(n) + \mu(n)\ x(n)\ \text{Sign}\{e(n)\} \tag{3.54}$$

$$w(n+1) = w(n) + \mu(n)\ \text{Sign}\{x(n)\}\ \text{Sign}\{e(n)\} \tag{3.55}$$

where $\mu(n) = \mu_{\max} + (\mu_{\min} - \mu_{\max})e^{-\beta}\sigma^2 e^{(n)}$ is the variable step size.

From (3.54) and (3.55), it can be noted that the weight update relations for SVSLMS and SSVSLMS are the same as SLMS and SSLMS algorithms.

3.10.3 Sign-Based Least Mean Logarithmic Squares (LMLS) Algorithms

LMLS algorithm is a basic version of higher-order adaptive filter. Combining the LMLS in (3.34) with SRA-, SA-, and SSA-based algorithms contributes to SRLMLS, SLMLS, and SSLMLS algorithms, respectively. Thus, the weight update expressions of signed variants of LMLS algorithms are given as follows:

$$w(n+1) = w(n) + \mu \cdot Sign\{x(n)\}\ e(n)\left[\frac{\alpha(e(n))^2}{1+\alpha(e(n))^2}\right] \tag{3.56}$$

$$w(n+1) = w(n) + \mu \cdot x(n)\text{Sign}\{e(n)\}\left[\frac{\alpha(e(n))^2}{1+\alpha(e(n))^2}\right] \tag{3.57}$$

$$w(n+1) = w(n) + \mu \cdot \text{Sign}\{x(n)\}\text{Sign}\left\{e(n)\left[\frac{\alpha(e(n))^2}{1+\alpha(e(n))^2}\right]\right\} \tag{3.58}$$

3.10.4 Sign-Based Least Logarithmic Absolute Difference (LLAD) Algorithms

LLAD algorithm is a basic version of higher-order adaptive filter. Combining the LLAD in (3.42) with SRA-, SA-, and SSA-based adaptive algorithms results in

SRLLAD, SLLAD, and SSLLAD algorithms, respectively. Therefore, the weight update relations of signed LLAD-based variants are given as follows:

$$w(n+1) = w(n) + \mu \cdot \text{Sign}\{x(n)\} \cdot \text{Sign}\left\{(e(n)).\left[\frac{\alpha(|e(n)|)}{1+\alpha(|e(n)|)}\right]\right\} \tag{3.59}$$

$$w(n+1) = w(n) + \mu \cdot x(n) \cdot \text{Sign}\left\{(e(n)).\left[\frac{\alpha(|e(n)|)}{1+\alpha(|e(n)|)}\right]\right\}, \tag{3.60}$$

$$w(n+1) = w(n) + \mu \cdot \text{Sign}\{x(n)\} \cdot \text{Sign}\left\{(e(n)).\left[\frac{\alpha(|e(n)|)}{1+\alpha(|e(n)|)}\right]\right\} \tag{3.61}$$

3.11 COMPUTATIONAL COMPLEXITY ISSUES

In general, the amount of multiply computations needed to finish operation is considered as a metric so that difficulty of the algorithms is compared and estimated. Yet, most DSPs have in-built MAC calculations. This procedure normally takes place in one instruction cycle and is added or subtracted. We focus on the comparative study of computational complexities among distinct LMS-based adaptive algorithms based on sign function.

Table 3.7 summarizes the computer complexity figures necessary to calculate using all LMS-based adaptive filters and their sign-based realizations. Also, since they remain chiefly free from multiplications, these techniques offer sophisticated ways to locate exon segments within a genomic sequence more accurately. For instance, the weight update equation in (3.17), when T is the length of the filter, is required to compute LMS algorithm with $T + 1$ multiplication operations and $T + 1$ addition operations. Only a single multiply operation is required for the SRA to compute $\mu \cdot e(n)$. Two other sign LMS algorithms do not involve multiplications, when μ is selected as a power of 2.

In such instances, multiply operation becomes low complicated in practice when it is changed into shift computation. Signum function is applied in SSA for both vector and data, and μ is added to the vector of weight with addition using sign check (ASC) computation.

From Table 3.7, it is evident that LMLS algorithm requires only four multiplications and zero addition operations, whereas LLAD algorithm requires only three multiplication operations with no additions. However, both LMLS and LLAD algorithms offer low computational complexity compared to all other LMS-based variants. Among these two algorithms, LLAD offers low complexity in performing computations and is more suitable in real bioinformatics-based applications.

To further reduce the computational complexity, we have considered signed variants of all LMS-based adaptive algorithms as shown in Table 3.7. It is clear that all their SRA-based techniques require lower multiply computations, because of the sign term applied to data vector, i.e. all the elements of the data vector become −1 or 0

TABLE 3.7
Computational Complexities of Various Sign LMS-Based Adaptive Techniques

S. No.	Algorithm	Multiplications	Additions	Divisions
1	LMS	T+1	T+1	Nil
2	SRLMS	1	T+1	Nil
3	SLMS	T	T+1	Nil
4	SSLMS	T+1	T+1	Nil
5	LMF	T+3	T+1	Nil
6	SRLMF	3	T+1	Nil
7	SLMF	T	T+1	Nil
8	SSLMS	T+1	T+1	Nil
9	VSLMS	T+1	T+1	Nil
10	SRVSLMS	1	T+1	Nil
11	SVSLMS	T	T+1	Nil
12	SSVSLMS	T+1	T+1	Nil
13	LMLS	T+7	T+2	1
14	SRLMLS	2T+7	2T+2	1
15	SLMLS	T+7	T+2	1
16	SSLMLS	T+2	T+2	1
17	LLAD	T+4	T+2	1
18	SRLLAD	4	T+2	1
19	SLLAD	T+3	T+2	1
20	SSLLAD	T+2	T+2	1

or 1, so that multiplications are greatly reduced. In the case of sign applied to error, there is no significant change in the number of multiplications, whereas in sign-sign-based algorithms, both data vector and error term use signed function, and therefore, no multiplication is needed to compute the weight update equation. All the computations can be carried out using addition operations with sign check.

From these computations, it is vibrant that SRLLLAD algorithm requires a lesser number of multiplications compared to all other signed LMS-based algorithms. Therefore, this algorithm-based AEP may be more appropriate for remote clinical health systems and bioinformatics systems. Because of clipping, only a single multiplication of a standard LMS in combination with SRA is required, while the SRLLAD requires less multiplication and is separate from filter duration. The logarithmic cost function and the sign function involved as part of weight relation remain responsible for the lower amount of four multiplications and one division needed for SRLLAD-based AEP. But the performance of the signal will degrade owing to clipping both data and error in the case of SSLLAD-based AEP. Therefore, in applications related to signal processing, SSLLAD could not be a better option. On the basis of these deliberations and less calculations, it is clear that SRLLAD might be the best choice for precise forecast of exons in a DNA sequence in GSP applications.

3.12 CONVERGENCE ANALYSIS

The convergence plots for distinct adaptive LMS algorithms mentioned in the previous sections are presented in Figure 3.9. They are drawn among a number of iterations and MSE. Calculation of MSE remains performed with the expression, i.e., $J = E\left[\left|e(n)^2\right|\right]$, in (3.8). The simulations for all the convergence curves are done using MATLAB. A discussion on the convergence analysis of stochastic gradient adaptive filters using the sign algorithm in literature is presented in Ref. [96]. In this section, calculation of the average value for the characterization for each sample and MSE is done for each sample in 4,000 iterations. The curves are achieved using different LMS-based AEPs separately with an adaptive filter with a finite length, a random variance of 0.01, and a step size of 0.01 for locating the exon segments. It is clear from the convergence plots of LMLS and LLAD that they outperform among all the sign-based LMS adaptive algorithms. Thus, the performances of all LMF, VSLMS, LMLS, and LLAD algorithms are analogous to each other and also superior to LMS.

Figures 3.10–3.14 show the convergence plots of all sign LMS-based algorithms. Due to the normalization factor, the performance of sign regressor version of all LMS algorithms is slightly inferior to the corresponding non-signed version of the algorithm; the corresponding convergence curves are shown in Figure 3.10. Due to its lower computational complexity, sign regressor LMS-based algorithm offers better exon tracking ability and is suitable for real-time applications. The LMF algorithm offers considerable advantages in reducing the signal distortion.

The SRLMF algorithm is evident to manage variations that occur within the PSD of exons during transmission. The parameter α is used to effectively

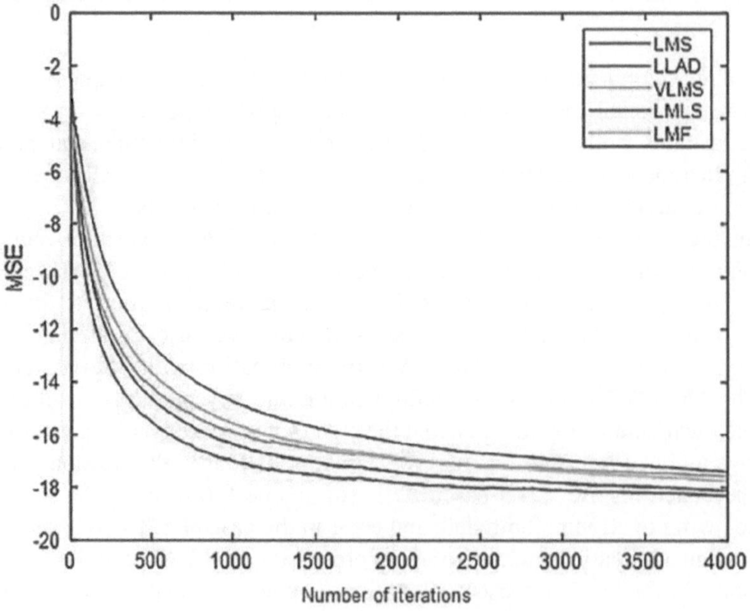

FIGURE 3.9 Convergence curves for LMS-based adaptive algorithms.

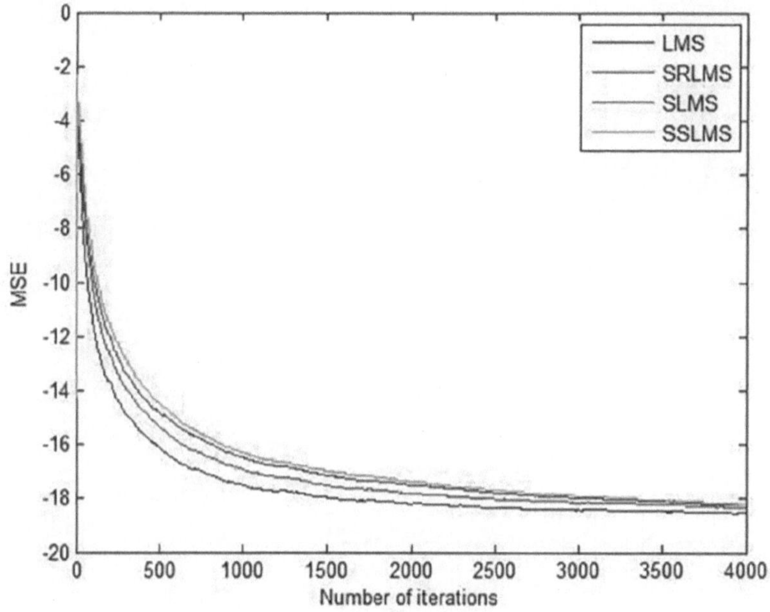

FIGURE 3.10 Convergence curves for sign-based LMS adaptive algorithms.

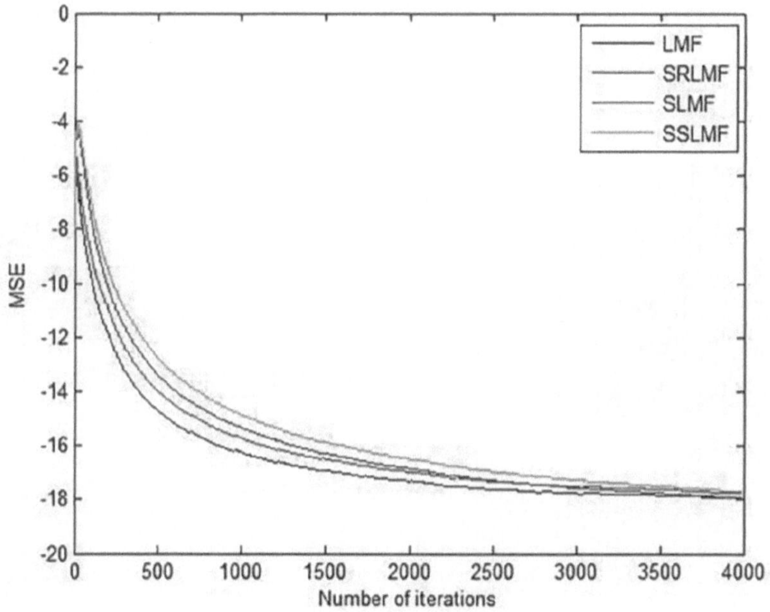

FIGURE 3.11 Convergence curves for sign-based LMF adaptive algorithms.

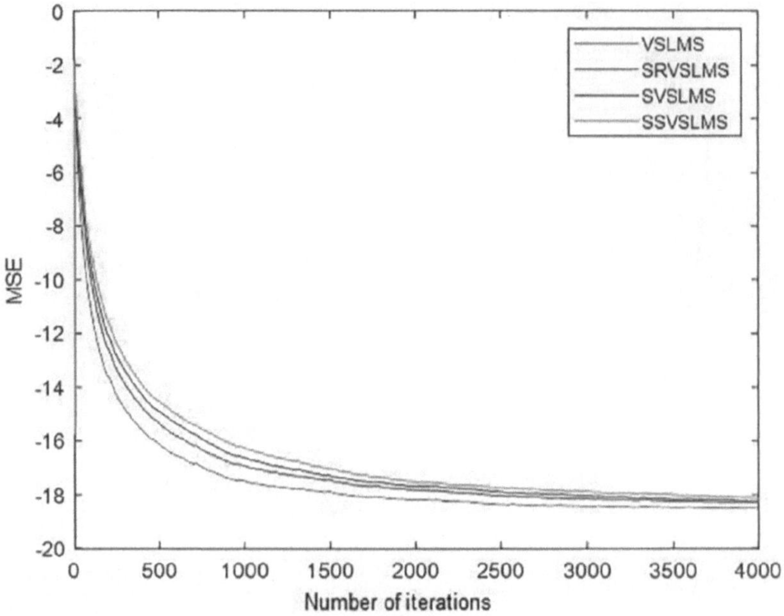

FIGURE 3.12 Convergence curves for sign-based VSLMS adaptive algorithms.

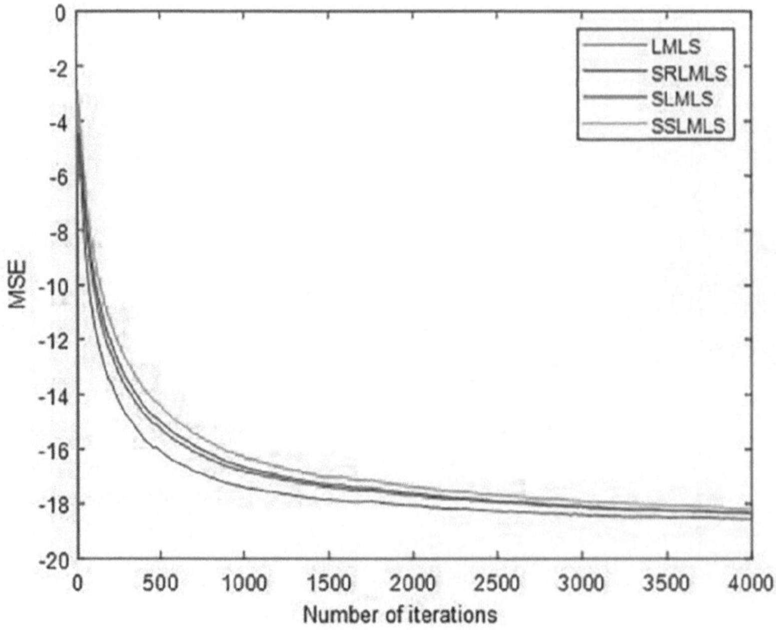

FIGURE 3.13 Convergence curves for sign-based LMLS adaptive algorithms.

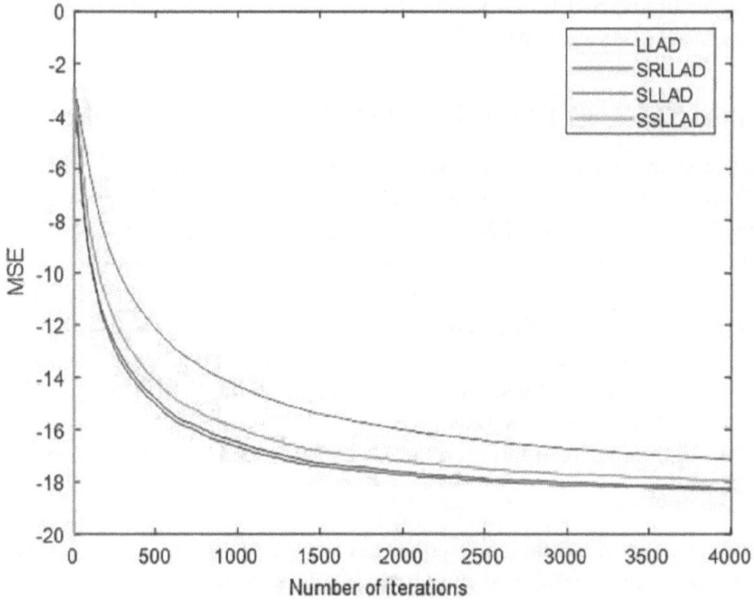

FIGURE 3.14 Convergence curves for sign-based LLAD adaptive algorithms.

control convergence. When a major error occurs, α tends towards unity and convergence becomes faster. Likewise, if α value is low, convergence will be slow with a small step size. In fact, this happens when the adaptive filter attains the steady state. To further improve convergence performance, VSLMS algorithm and its signed variants are considered. The convergence equations are obtained through multiple samples of MSE. MSE was noted to decrease over iterations and samples besides specifically over change in the value of α. This demonstrates that with increase in its value, error term remains weighted more, thus efficiently suppressing the noise. From these features, it is evident that all suggested LMLS-based adaptive techniques offer a quicker convergence rate compared to LMS.

Therefore, among different algorithms regarded for implementing AEPs, SRLMLS-based adaptive AEP is preferred as a better candidate due to its higher convergence performance and lower computational difficulty than other algorithms.

By implementing signum function, however, convergence becomes delayed. The hybrid variant of SRA is just lower than its non-signed variant, according to the illustrations. Figure 3.14 shows that all the LLAD-based adaptive algorithms have a higher rate of convergence than LMS. Therefore, the SRLLAD-based adaptive algorithm, compared to all other LMS-based algorithms, is considered to be the best of all other variants, when LLAD was considered for the implementation of AEPs. The signed algorithms will always exhibit sub-optimal convergence performance when compared to their conventional realizations, since they tend to reject some information in exchange for lower complexity.

The same phenomenon is observed for all signed regressor versions of LMF, VSLMS, LMLS, and LLAD algorithms. Because of its faster convergence and better predictive accuracy, SRLLAD-based AEP can be used in practice among all sign regressor–based LMS variants.

3.13 RESULTS AND DISCUSSION FOR LMS-BASED VARIANTS

This section compares the performance of different LMS-based AEPs in this chapter. AEP structure is as demonstrated in Figure 3.2. Various LMS-based algorithms including LMF, VSLMS, LMLS, LLAD, and their sign-based versions of different AEPs are used for implementation. AEP using LMS technique is also created for the purpose of comparative study. For the purpose of evaluation, ten genomic sequences were obtained from the NCBI databank [40]. In order to assess the efficiency of different algorithms, the results were consistent; we perceived these sequences to be our set of genomic data. Several gene sequences taken from the NCBI gene databank and their descriptions are shown in Table 3.8.

The flow diagram shown in Figure 3.15 with steps for adaptive exon prediction is described below:

a. The selection of a gene input sequence from the NCBI databank uses a new adaptive exon identification method. Input DNA sequence is then evaluated on the basis of A+T along with G+C base pair nucleotide densities with density plots for gene locations. After the assessment, conversion of gene sequence to digital form is done using a digital mapping method. Figure 3.15 shows the AEP input as binary data.
b. The resultant sequence can now be given as an AEP input. The biological sequence that obeys to TBP property is provided as a reference signal for AEP.
c. For updating filter coefficients, the signal for the feedback signal $e(n)$ is extracted from Figure 3.2.
d. When this signal is minimized, exon sections are positioned correctly with the PSD plot from the DNA sequence.

TABLE 3.8
Gene Datasets from NCBI GenBank Database

Seq. No.	Accession No.	Sequence Definition
1	E15270.1	Human gene for osteoclastogenesis inhibitory factor (OCIF) gene
2	X77471.1	Homo sapiens human tyrosine aminotransferase (TAT) gene
3	AB035346.2	Homo sapiens T-cell leukemia/lymphoma 6 (TCL6) gene
4	AJ225085.1	*Homo sapiens* Fanconi anemia group A (FAA) gene
5	AF009962	*Homo sapiens* CC-chemokine receptor (CCR-5) gene
6	X59065.1	*Homo sapiens* human acidic fibroblast growth factor (FGF) gene
7	AJ223321.1	*Homo sapiens* transcriptional repressor (RP58) gene
8	X92412.1	*Homo sapiens* titin (TTN) gene
9	U01317.1	Human beta globin sequence on chromosome 11
10	X51502.1	*Homo sapiens* gene for prolactin-inducible protein (GPIPI)

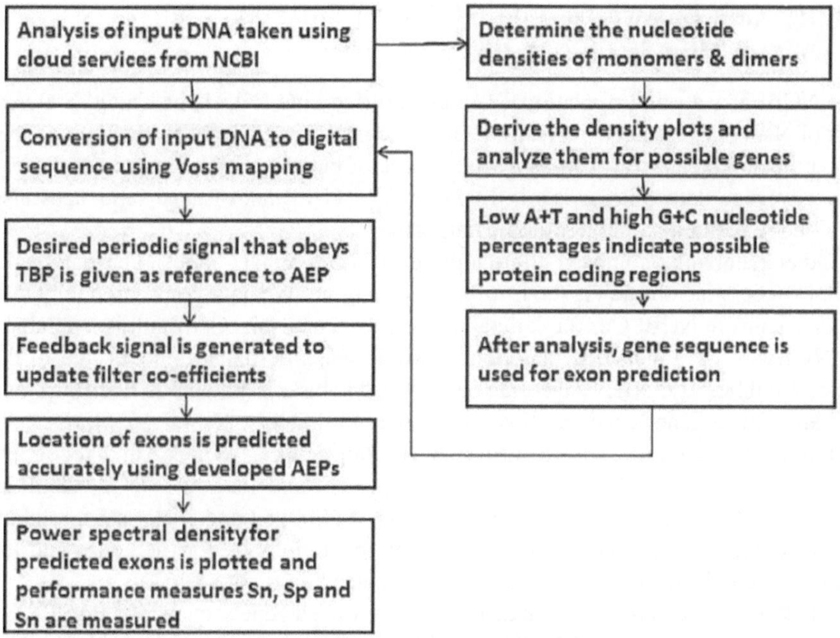

FIGURE 3.15 Flow diagram for adaptive exon prediction.

e. The plots for located exon sections are expressed with PSD. In addition, Sn, Pr, and Sp are derived and plotted.

Here, we present various LMS-based implementation of adaptive filtering algorithms to locate exon sections accurately in the input gene sequence of *Homo sapiens*. We developed AEPs that branch from LMS algorithm. This is the first category of the adaptive algorithms we have considered. In this category, we developed five types of AEPs based on the LMS-based algorithms that include LMS, LMF, VSLMS, LMLS, and LLAD algorithms. As in case of larger sequences, the samples overlap at the exon predictor input on each other and hence less computational complex adaptive filter is desirable. To achieve this, we extend the above adaptive filters with their sign-based realizations, which are known to be simpler than their LMS-based counterparts, requiring only 50% as many multipliers.

Three different versions of the sign-based algorithms are considered for this purpose, namely, (i) SRA, (ii) SA, and (iii) SSA techniques. Use of signum function in all the algorithms requires appropriate modifications in the weight update recursion. Using all these combinations, various AEPs are being created and tested, and different real DNA sequences of *Homo sapiens* taken from NCBI gene databank are presented. It is evident from the PSD plots of exon prediction that signed regressor variants of all LMS-based AEPs developed using LMF, VSLMS, LMLS, and LLAD algorithms predicts the exon locations with more accuracy compared to LMS and other variants.

3.13.1 Gene Datasets from the NCBI Gene Databank for Gene Sequence Analysis

The NCBI is a portion of National Library of Medicine (NLM) of United States, a unit of National Institutes of Health (NIH). It contains a group of biotechnology and biomedical databanks as well as a vital source of bioinformatics services and tools. Majority gene databanks exists in literature like GenBank of gene sequences also PubMed, a biomedical bibliographic databank.

Other databanks comprise the database of NCBI Epigenomics. The databanks remain accessible online via the Entrez search engine. We have considered ten DNA sequences from NCBI GenBank database for DNA analysis. GenBank is a database by NIH with an annotated collection of whole genomic datasets accessible to the general public. The NIH nucleotide sequence database is known as GenBank and contains whole gene sequences that are publicly available.

For instance, one can go to http://www.ncbi.nlm.nih.gov/entrez and recognize accession number AF009962 of a DNA sequence, and then choose 'Nucleotide' under Search and fill the other entry by typing: AF009962 and click on Search button. Clicking on the input sequence indicates the annotation for the genes and the entire nucleotide sequence in the form of raw data as in Figure 3.16.

Clicking on the "FASTA" link displays the retrieved gene sequence in FASTA file format as shown in Figure 3.16. The resulting sequence is to be saved with .FASTA file extension and use it for our DNA sequence analysis. Ten DNA sequence datasets considered from the GenBank gene database for our analysis for identifying the exon locations are presented in Table 3.3. Gene datasets considered from NCBI GenBank database in our current research are with accessions E15270.1, X77471.1, AB035346.2, AJ225085.1, AF009962, X59065.1, AJ223321.1, X92412.1, U01317.1, and X51502.1, respectively.

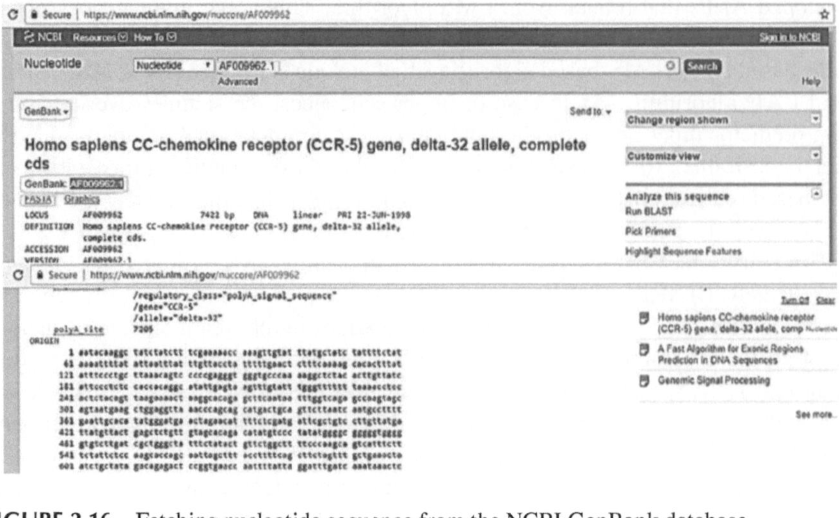

FIGURE 3.16 Fetching nucleotide sequence from the NCBI GenBank database.

3.13.2 Analysis of Gene Datasets of NCBI Gene Databank

3.13.2.1 Nucleotide Densities of Monomers and Dimers in Gene Dataset

DNA sections using an elevated percentage of A+T nucleotides generally show an intergenic sequence part, whereas small A+T and greater G+C nucleotide percentages specify likely genes. Elevated content of CG dinucleotides is often found before a gene. Using the sequence statistics function, one can determine if gene sequence contains the features of a protein-coding region when reading into the MATLAB environment.

The human genomic sequence with accession AF009962 is used for this operation. Figure 3.17 illustrates the typical plot of monomer and dimer nucleotide densities with AF009962 accession for the nucleotide sequence. It is evident from the above plot that monomer densities for T nucleotide bases are present along the length of the sequence and its density is more than other nucleotide bases A, C, and G in a gene sequence with accession AF009962. Figure 3.18 shows a typical pie illustration for the monomer base counts distribution in a nucleotide sequence with accession AF009962. A list of 5'–3' strand nucleotide base counts is shown in a DNA sequence with accession AF009962 below to visualize the nucleotide distribution.

In this gene sequence, there are 2126 bases of A nucleotide, 1477 monomer bases of C, 1667 bases of G, and 2152 nucleotide bases of T are available as depicted in

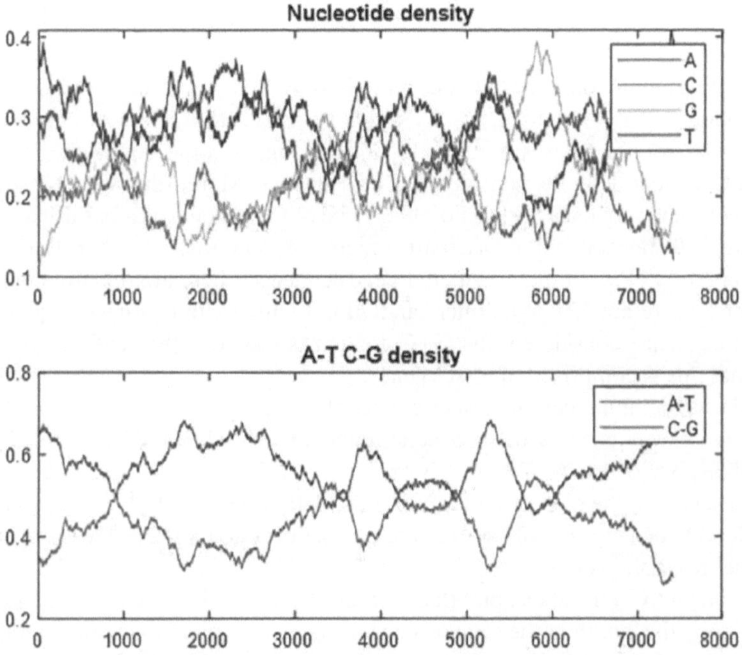

FIGURE 3.17 Monomer and dimer nucleotide densities for a human genome sequence with accession AF009962.

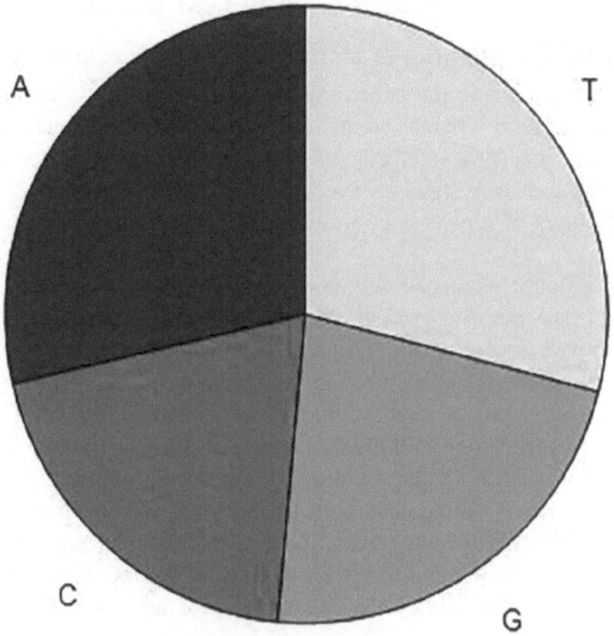

FIGURE 3.18 Monomer base counts distribution in a gene sequence with accession AF009962.

Figure 3.18. The monomer base count for T is slightly lesser than the base count of A nucleotide.

Also, the monomer bases of G nucleotide are higher compared to the monomer bases of C nucleotide in the considered gene sequence. Moreover, the distribution of dimers in a gene sequence with AF009962 accession is shown as a bar illustration.

Figure 3.19 shows that T-T base pair dimers are more than all other dimers with accession AF009962. In the considered gene sequence, there are 680 dimers of T-T base pairs. There are 527 A-T dimers and 70 G-C dimers in this gene sequence. In the gene sequence considered, the G+C content is less than the A+T dimer, which shows that this is comprised of lesser genes.

Similarly, the monomer and dimer counts of nucleotide bases present in all other *Homo sapiens* sequences with accession numbers E15270.1, X77471.1, AB035346.2, AJ225085.1, X59065.1, AJ223321.1, X92412.1, U01317.1, and X51502.1 are derived and represented in the plots from Figures 3.20–3.46. Figure 3.20 represents the typical nucleotide density plot of the sequence with accession E15270.1 with monomer and dimer nucleotide densities.

It is evident from the above plot that monomer densities for T nucleotide bases are present along the length of the sequence and its density is more than other nucleotide bases A, C, and G in the gene sequence with accession E15270.1. Figure 3.21 shows a typical pie illustration for the monomer base counts distribution in a nucleotide sequence with accession E15270.1. A list of 5'-3' strand baseline counts displayed in

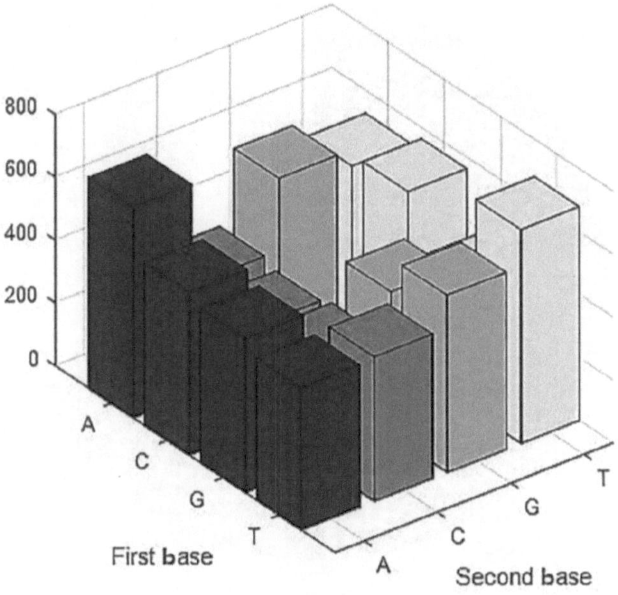

FIGURE 3.19 Dimer base counts distribution in a gene sequence with accession AF009962.

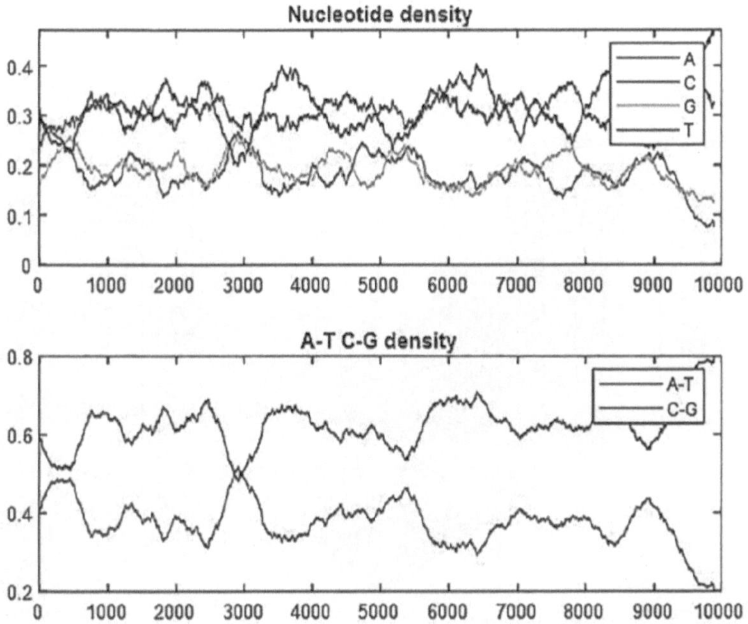

FIGURE 3.20 Monomer and dimer nucleotide densities for a human genome sequence with accession E15270.1.

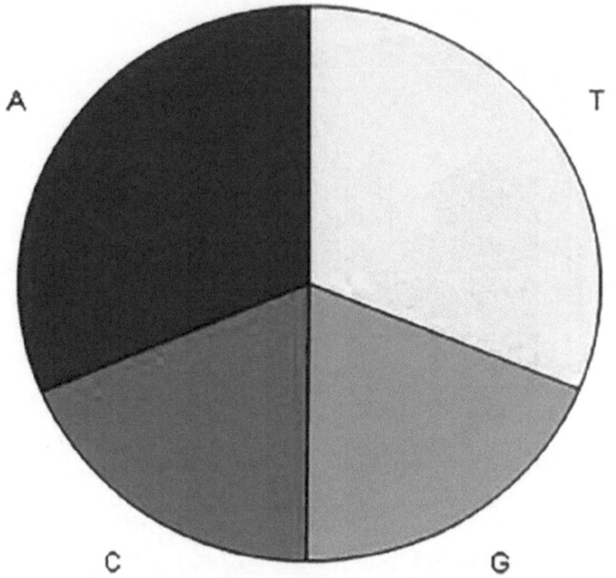

FIGURE 3.21 Monomer base counts distribution in a gene sequence with accession E15270.1.

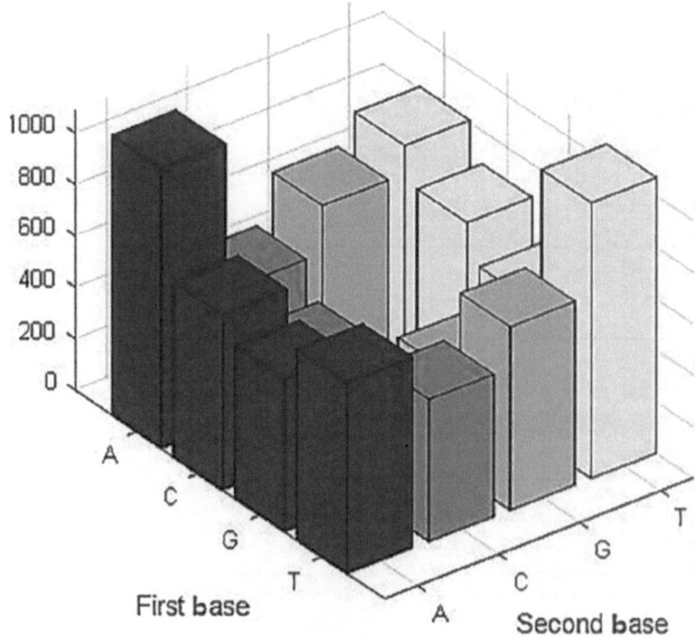

FIGURE 3.22 Dimer base counts distribution in a gene sequence with accession E15270.1.

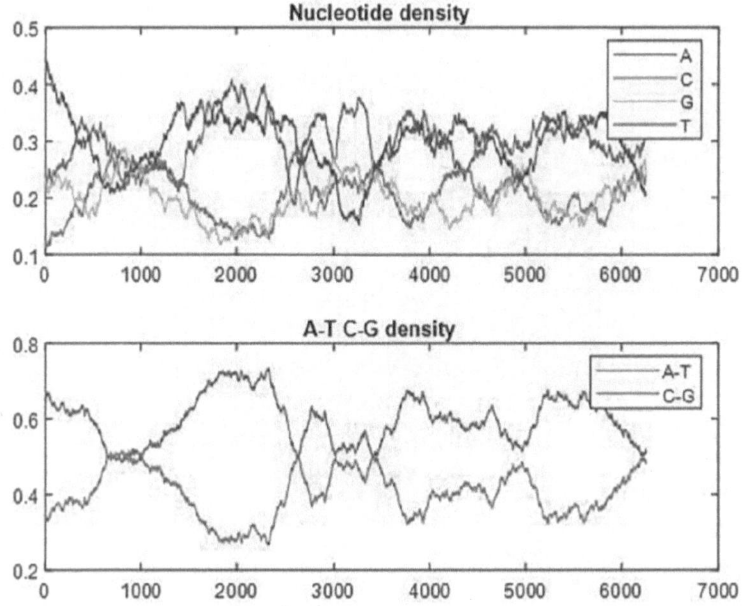

FIGURE 3.23 Monomer and dimer nucleotide densities for a human genome sequence with accession X77471.1.

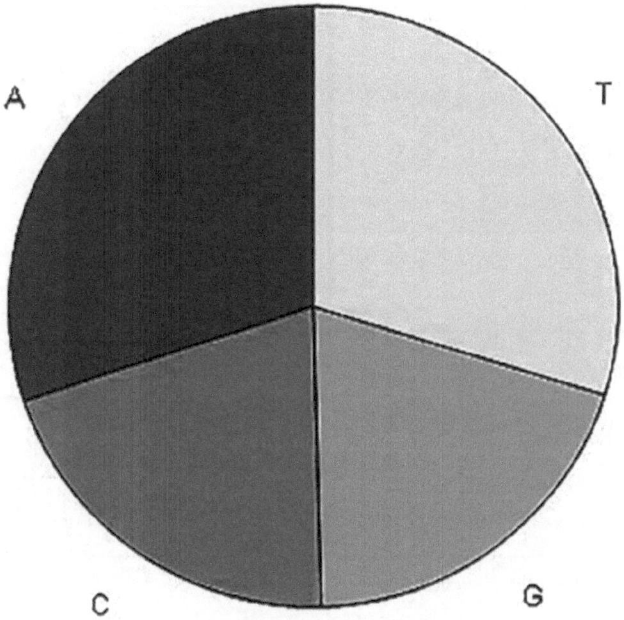

FIGURE 3.24 Monomer base counts distribution in a gene sequence with accession X77471.1.

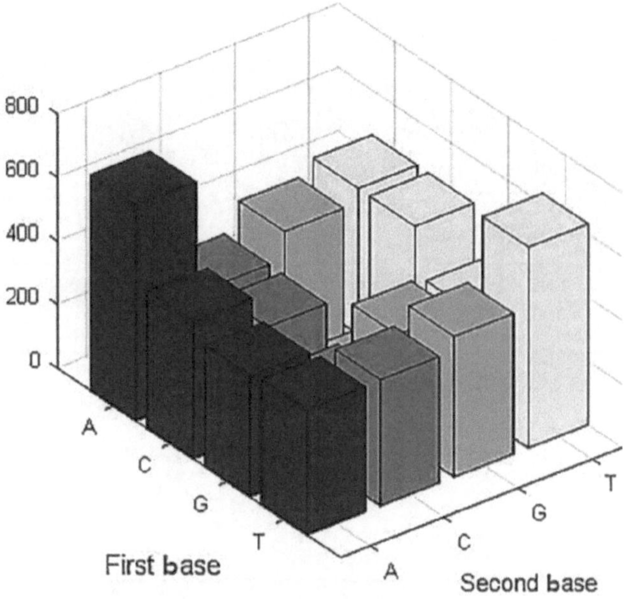

FIGURE 3.25 Dimer base counts distribution in a gene sequence with accession X77471.1.

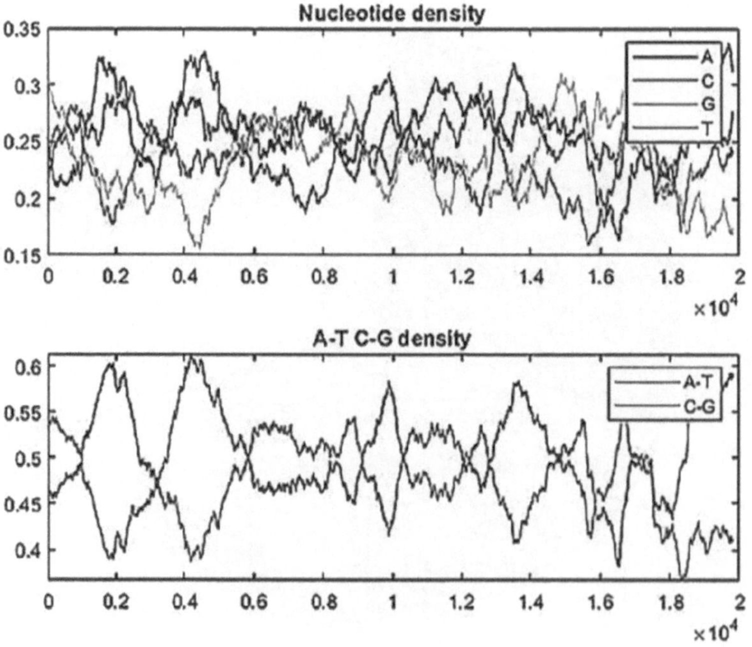

FIGURE 3.26 Monomer and dimer nucleotide densities for a human genome sequence with accession AB035346.2.

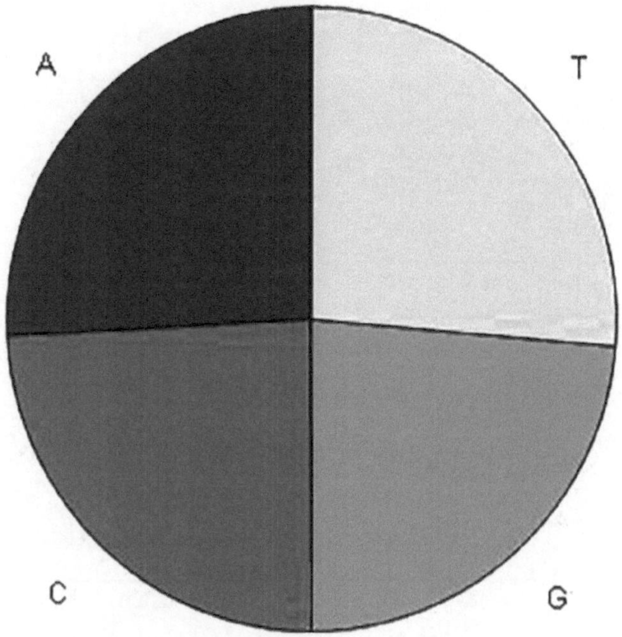

FIGURE 3.27 Monomer base counts distribution in a gene sequence with accession AB035346.2.

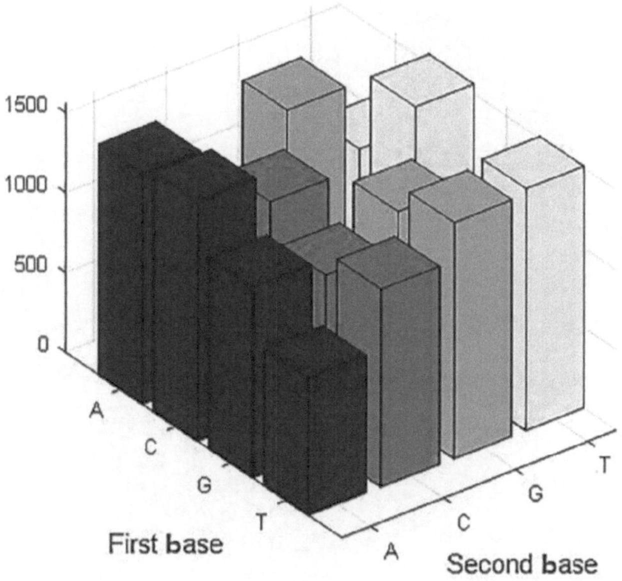

FIGURE 3.28 Dimer base counts distribution in a gene sequence with accession AB035346.2.

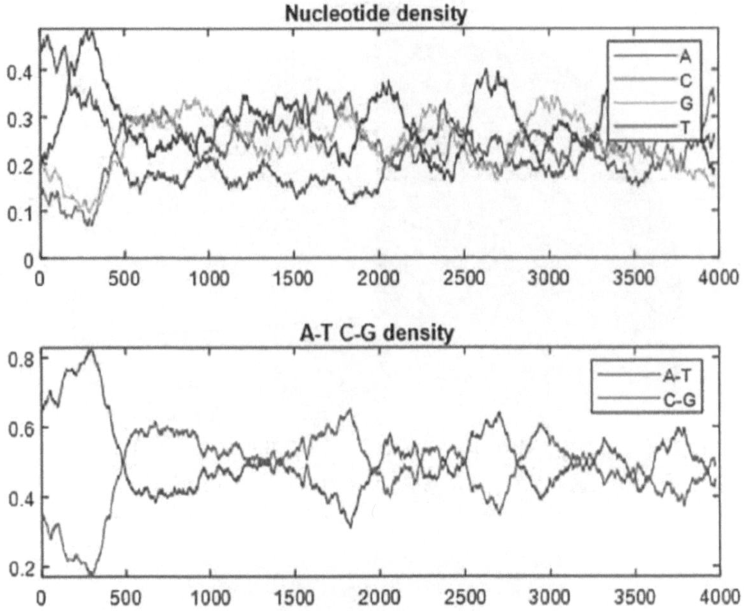

FIGURE 3.29 Monomer and dimer nucleotide densities for a human genome sequence with accession AJ225085.1.

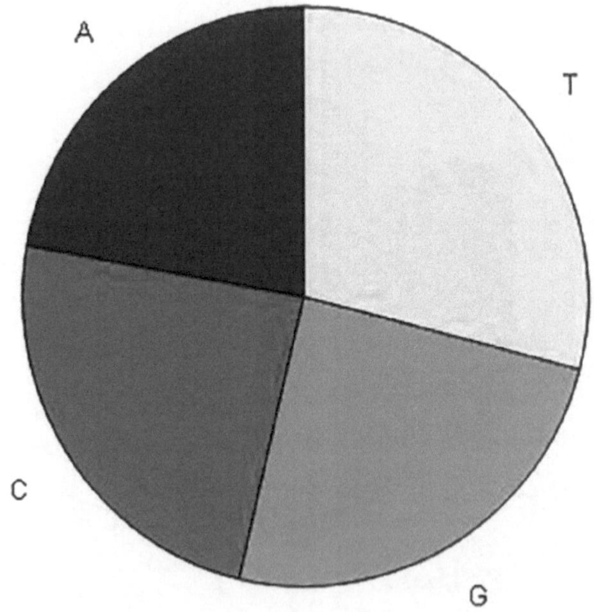

FIGURE 3.30 Monomer base counts distribution in a gene sequence with accession AJ225085.1.

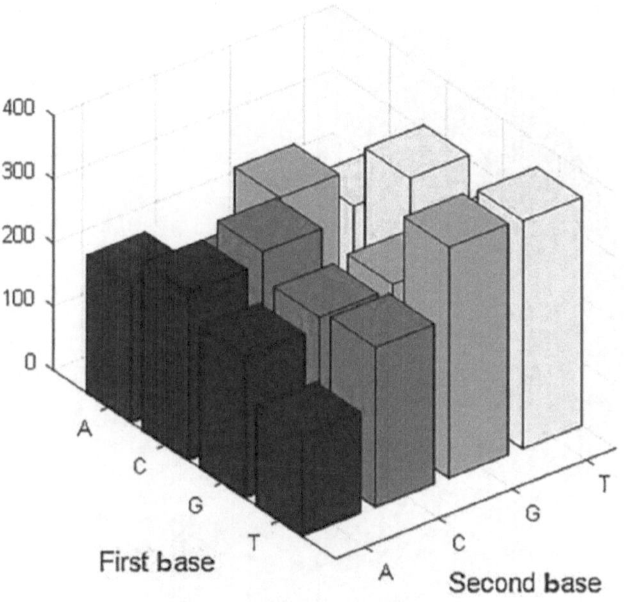

FIGURE 3.31 Dimer base counts distribution in a gene sequence with accession AJ225085.1.

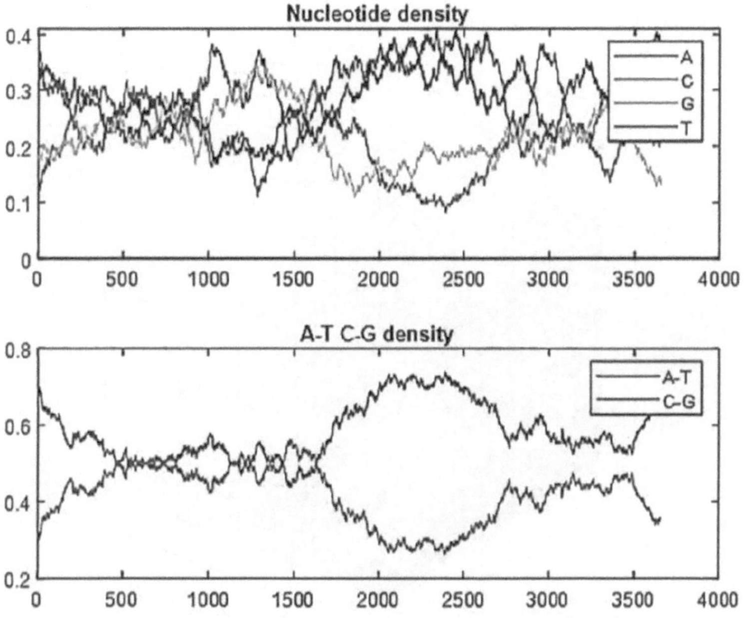

FIGURE 3.32 Monomer and dimer nucleotide densities for a human genome sequence with accession X59065.1.

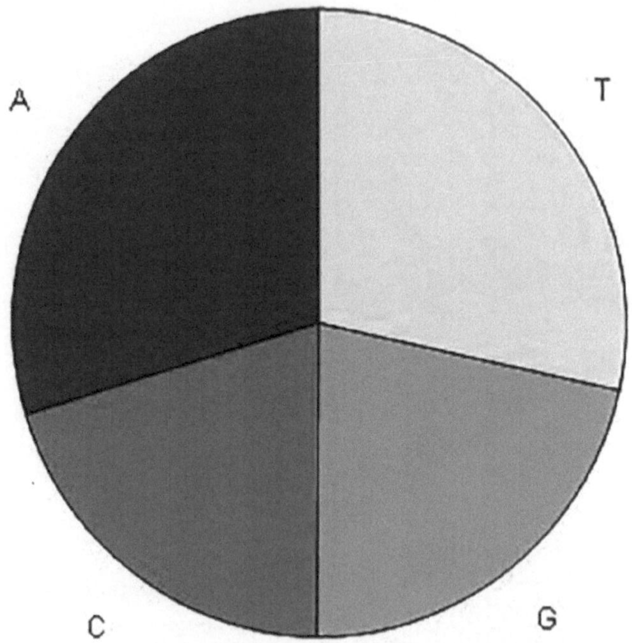

FIGURE 3.33 Monomer base counts distribution in a gene sequence with accession X59065.1.

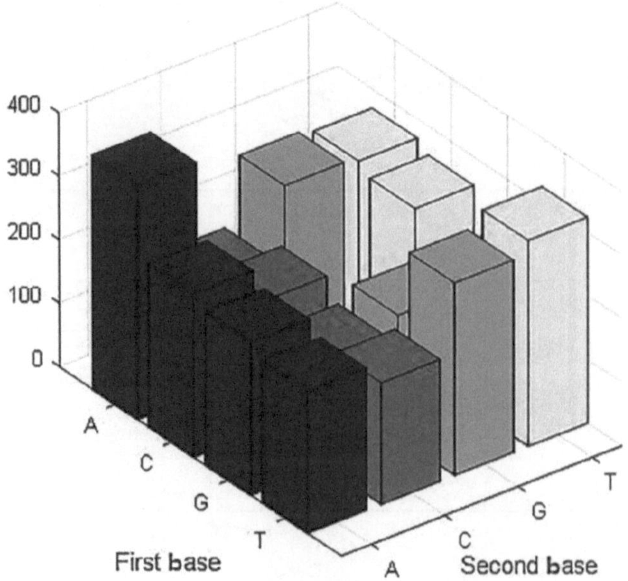

FIGURE 3.34 Dimer base counts distribution in a gene sequence with accession X59065.1.

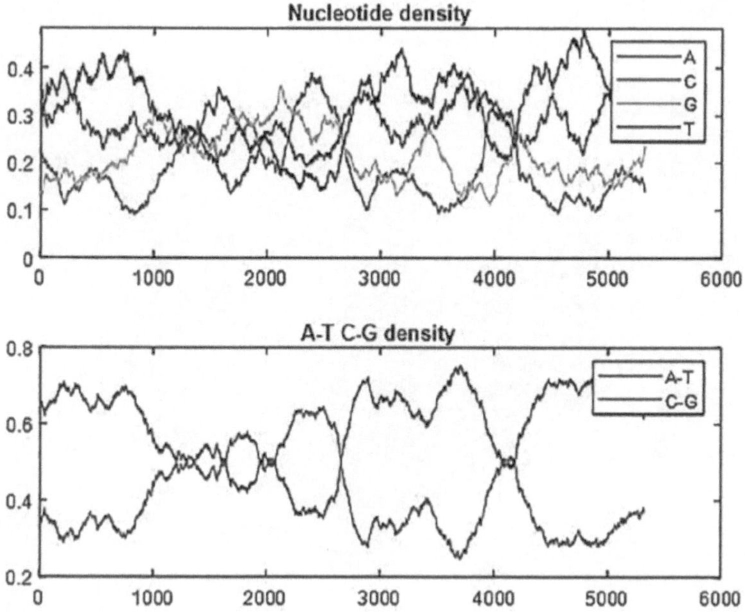

FIGURE 3.35 Monomer and dimer nucleotide densities for a human genome sequence with accession AJ223321.1.

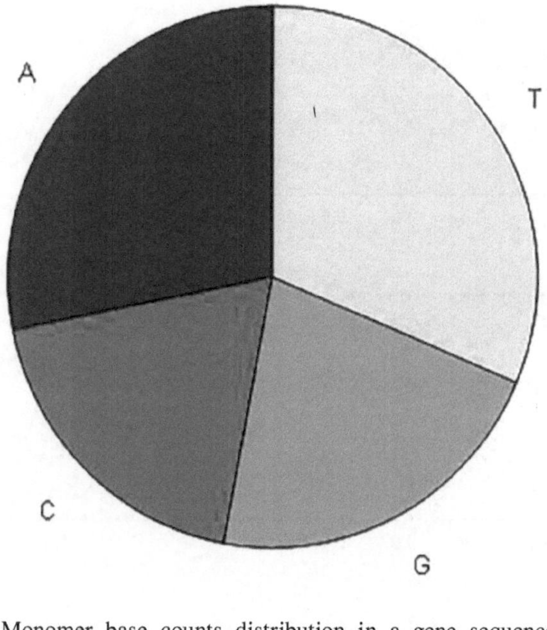

FIGURE 3.36 Monomer base counts distribution in a gene sequence with accession AJ223321.1.

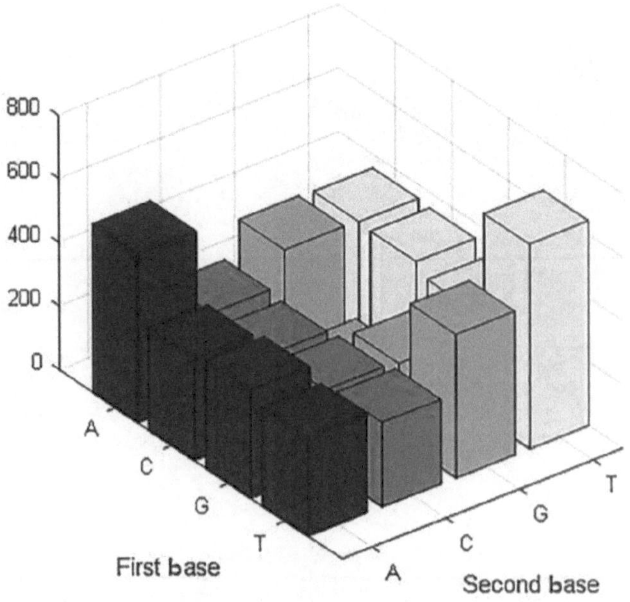

FIGURE 3.37 Dimer base counts distribution in a gene sequence with accession AJ223321.1.

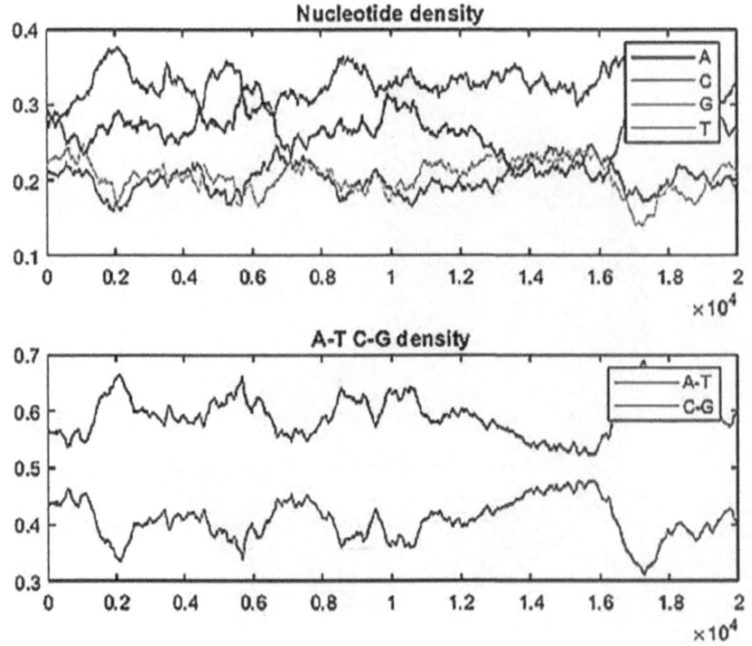

FIGURE 3.38 Monomer and dimer nucleotide densities for a human genome sequence with accession X92412.1.

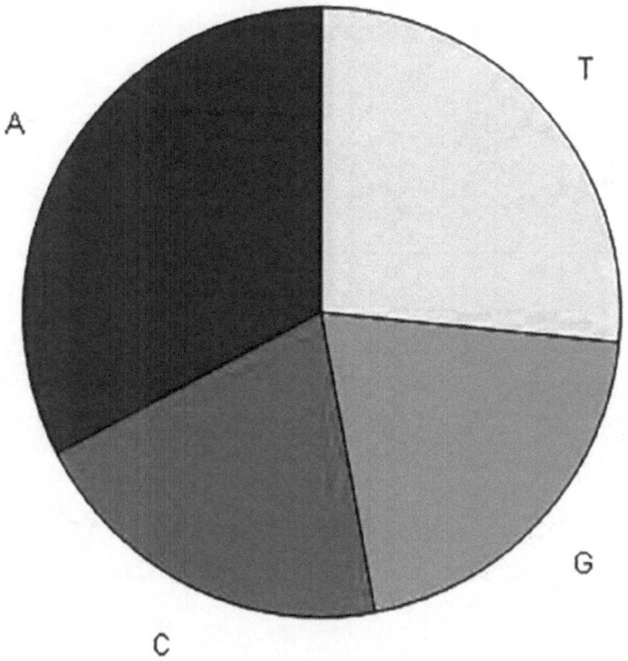

FIGURE 3.39 Monomer base counts distribution in a gene sequence with accession X92412.1.

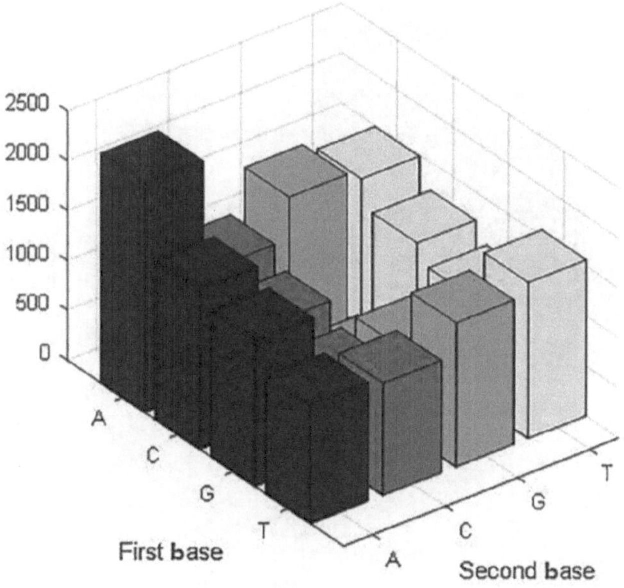

FIGURE 3.40 Dimer base counts distribution in a gene sequence with accession X92412.1.

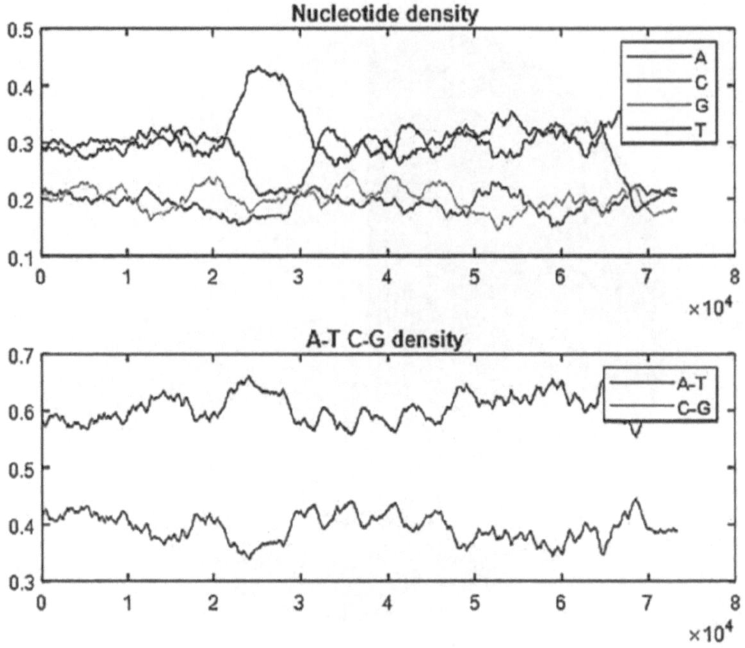

FIGURE 3.41 Monomer and dimer nucleotide densities for a human genome sequence with accession U01317.1.

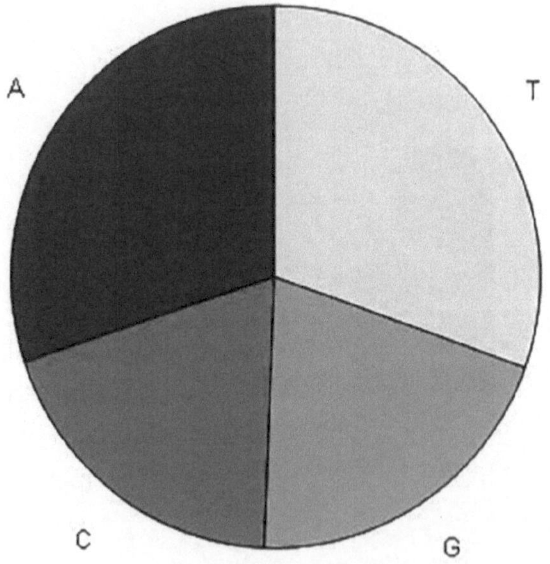

FIGURE 3.42 Monomer base counts distribution in a gene sequence with accession U01317.1.

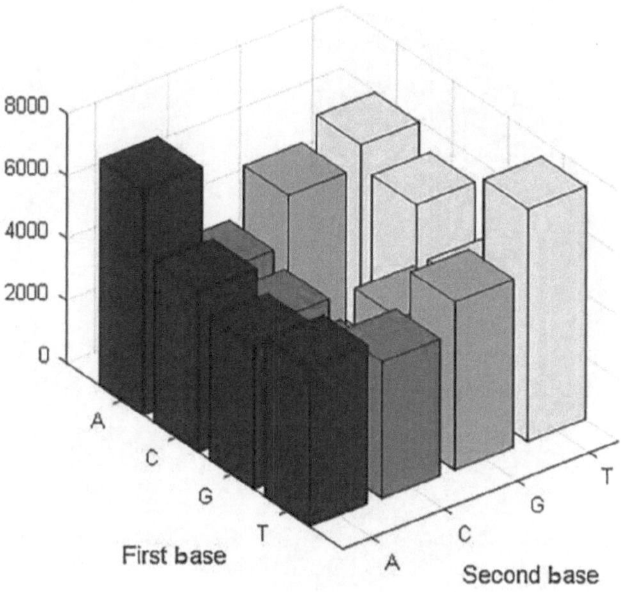

FIGURE 3.43 Dimer base counts distribution in a gene sequence with accession U01317.1.

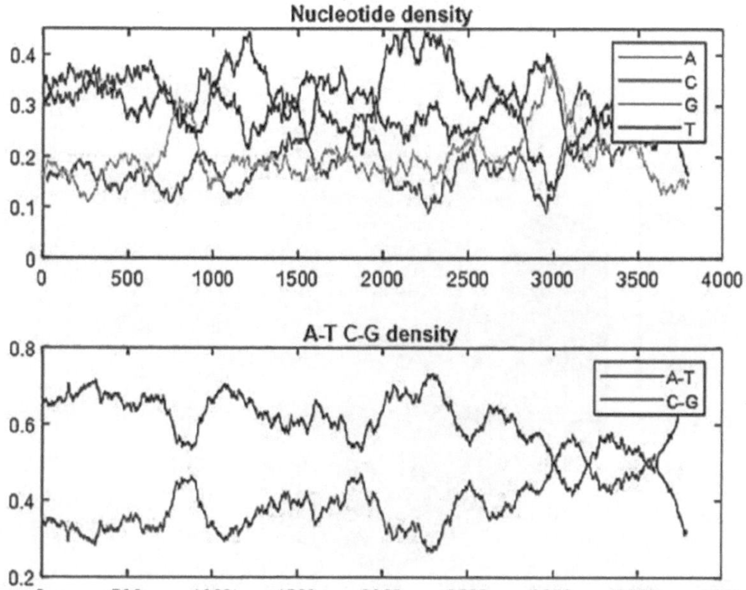

FIGURE 3.44 Monomer and dimer nucleotide densities for a human genome sequence with accession X51502.1.

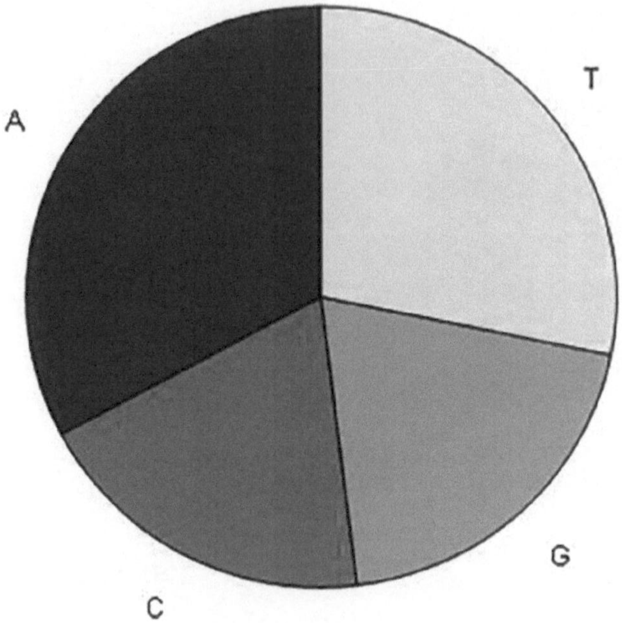

FIGURE 3.45 Monomer base counts distribution in a gene sequence with accession X51502.1.

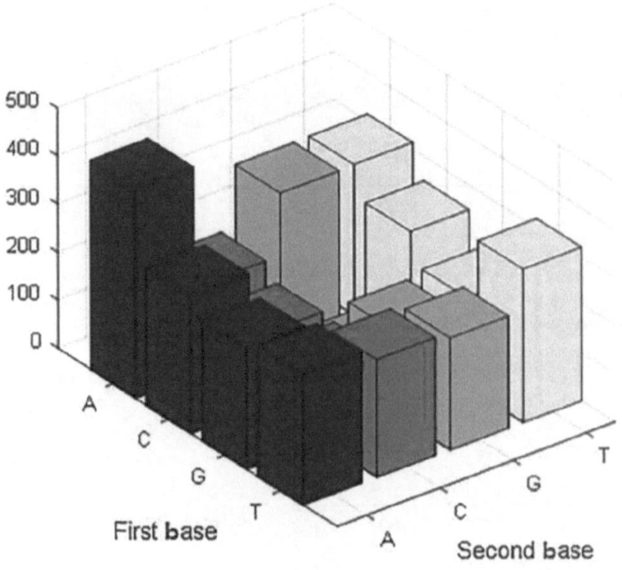

FIGURE 3.46 Dimer base counts distribution in a gene sequence with accession X51502.1.

a DNA sequence with accession E15270.1 is plotted below to visualize the nucleotide distribution.

In this gene sequence, 3110 bases of A nucleotide, 1825 monomer bases of C, 1881 bases of G, and 3082 nucleotide bases of T are available as depicted in Figure 3.21. Also, the monomer bases of G nucleotide are higher compared to monomer bases of C nucleotide in the considered gene sequence. In addition, the dimers are shown in the form of a bar illustration in a gene sequence with accession E15270.1. It is shown in Figure 3.22 that dimers using T-T base pairs are more compared to all other dimers in a gene sequence with accession E15270.1. In the gene sequence considered, there are 1072 dimers of T-T base pairs, 808 A-T dimers, and 377 G-C dimers. In the gene sequence considered, G+C content is less than A+T dimers, which shows that it has less genes. Figure 3.23 shows the typical plot with accession X77471.1 for monomer and dimer nucleotide densities in nucleotide sequence. It is evident from the above plot that monomer densities for T nucleotide bases are present along the length of the sequence and its density is more than other nucleotide bases A, C, and G in the gene sequence with accession X77471.1.

Figure 3.24 shows a typical pie illustration for the monomer base counts distribution in a nucleotide sequence with accession X77471.1. The 5'–3' strand shows a list of nucleotide stats in a DNA sequence with accession X77471.1 as plotted below to analyze the nucleotide distribution. In this gene sequence, 1882 bases of A nucleotide, 1271 monomer bases of C, 1239 bases of G, and 1861 nucleotide bases of T are available as depicted in Figure 3.24. The monomer base count for T is slightly lesser than the base count of A nucleotide. Also, the monomer bases of G nucleotide are higher compared to the monomer bases of C nucleotide in the considered gene sequence. The distribution of dimers in a gene sequence with accession X77471.1 is also shown in a bar illustration. It is shown in Figure 3.25 that dimers using T-T base pairs are more compared to all other dimers in a gene sequence with accession X77471.1. In the considered gene sequence, there are 679 dimers of A-A base pairs.

There are 459 A-T dimers and 221 G-C dimers in this gene sequence. The G+C content is lesser compared to A+T dimers in this sequence with accession X77471.1, which suggests that it has less genes. Figure 3.26 shows the typical plot with accession AB035346.2 for monomer and dimer nucleotide densities for nucleotide sequence. It is evident from the above plot that monomer densities for T nucleotide bases are present along the length of the sequence and its density is more than other nucleotide bases A, C, and G in the gene sequence with accession AB035346.2.

Figure 3.27 shows a typical pie illustration for the monomer base counts distribution in a nucleotide sequence with accession AB035346.2. A list of base nucleotide counts for 5'–3' strand in this DNA sequence is shown below to visualize the nucleotide distribution. In the gene sequence with accession AB035346.2, 5029 bases of A nucleotide, 4693 monomer bases of C, 4579 bases of G, and 5119 nucleotide bases of T are available as depicted in Figure 3.27.

The monomer base count for T is slightly lesser than the base count for A nucleotide. Also, the monomer bases of C nucleotide are higher compared to the monomer bases of G nucleotide in the considered gene sequence. The dimers' distribution in a gene sequence with accession AB035346.2 is also portrayed as a bar illustration. Figure 3.28 shows that dimers using T-T base pairs compared to all other dimers in the gene sequence with accession AB035346.2. This gene sequence has the highest base counts for C-T base pair dimers. In this gene sequence, there are 1066 A-T dimers and 1093 G-C dimers. In this sequence, G+C content is more compared to A+T dimers with accession AB035346.2 which indicates the presence of more genes. Figure 3.29 shows the typical plot for monomer and dimer nucleotide densities with accession AJ225085.1. Also, Figure 3.30 shows a typical pie illustration for the monomer base counts distribution in a nucleotide sequence with accession AJ225085.1.

The 5'–3' strand displays a list of nucleotide base counts in a DNA sequence with accession AJ225085.1 as plotted below to visualize the nucleotide distribution. There are gene sequences with accession AJ225085.1 has 869 bases of A nucleotide, 947 monomer bases of C, 967 bases of G and 1140 nucleotide bases of T are available as depicted in the Figure 3.30.

The monomer base count of C is slightly lesser than the base count of G nucleotide. Also, the monomer bases of T nucleotide are higher compared to the monomer bases of G nucleotide in the considered gene sequence. The distribution of the dimer in a gene sequence with accession AJ225085.1 is also shown as a bar illustration.

Figure 3.31 shows that dimers with T-T base pairs are more important than all other dimers within the gene sequence with accession AJ225085.1. Maximum base counts are available in this gene sequence with 363 dimers of T-G base pairs. There are 204 A-T dimers and 239 G-C dimers in this gene sequence. In this sequence, G+C content is higher than A+T dimers with accession AJ225085.1, indicating that it has more genes. The typical plot for monomer and dimer nucleotide densities for the X59065.1 nucleotide accession sequence is shown in Figure 3.32.

It is evident from the above plot that monomer densities for T nucleotide bases are present along the length of sequence and its density is more than other nucleotide bases A, C, and G in the gene sequence with accession X59065.1. Figure 3.33 shows a typical pie illustration for the monomer base counts distribution in a nucleotide sequence with accession X59065.1.

A list of the nucleotide base counts for the 5'–3' strand in a gene sequence with accession X59065.1 is shown below to present the nucleotide distribution. There are gene sequences with accession X59065.1 has 1088 bases of A nucleotide, 737 monomer bases of C, 793 bases of G and 1040 nucleotide bases of T are available as depicted. The monomer base count of T is slightly lesser than the base count of A nucleotide. Also, the monomer bases of G nucleotide are higher compared to the monomer bases of C nucleotide in the considered gene sequence.

In addition, the distribution of dimers in a gene sequence with accession X59065.1 is illustrated in a bar plot. Figure 3.34 shows that dimers with T-T basis pairs are more common in the gene sequence with entry X59065.1 than all other dimers. In this gene sequence, there are maximum base counts with 368 dimers of A-A base pairs, 270 A-T dimers, and 171 G-C dimers.

G+C content is less than A+T dimers in this sequence with X59065.1 accession, which shows that it has less genes. Figure 3.35 shows the typical plot for the monomer and dimer densities of a nucleotide sequence with accession AJ223321.1. It is evident from the above plot that monomer densities for T nucleotide bases are present along the length of sequence and its density is more than other nucleotide bases A, C, and G in a gene sequence with accession AJ223321.1.

Figure 3.36 shows a typical pie illustration for the monomer base counts distribution in a nucleotide sequence with accession AJ223321.1. The 5'–3' strand shows a list of nucleotide base counts in a DNA sequence with accession AJ223321.1 as plotted below to analyze the nucleotide distribution. In the gene sequence with accession AJ223321.1, 1,500 bases of A nucleotide, 1,007 monomer bases of C, 1,146 bases of G, and 1,668 nucleotide bases of T are available as depicted in Figure 3.36.

The monomer base count for T is slightly lesser than the base count for A nucleotide. Also, the monomer bases of G nucleotide are higher compared to the monomer bases of C nucleotide in the considered gene sequence. The distribution of dimers in the gene sequence with accession AJ223321.1 is also represented as a bar illustration.

Figure 3.37 shows that T-T-base pair dimers are higher compared with all other dimers within the gene sequence with accession AJ223321.1. Maximum base counts with 643 T-T base pairs in this gene sequence are available.

There are 357 A-T dimers and 249 G-C dimers in this gene sequence. Compared to the A+T dimers, G+C content is less in the sequence with accession X92412.1. Figure 3.38 shows the typical plot with accession X92412.1 for monomer and dimer nucleotide densities of the nucleotide sequence. It is evident from the above plot that monomer densities for T nucleotide bases are present along the length of the sequence and its density is more than other nucleotide bases A, C, and G in the gene sequence with accession X92412.1. Figure 3.39 shows a typical pie illustration for the monomer base count distribution in a nucleotide sequence with accession X92412.1. The 5'–3' strand displays a list of nucleotide base counts in a DNA sequence with accession X92412.1 as plotted below to visualize the nucleotide distribution.

In the gene sequence with accession X92412.1, 6535 bases of A nucleotide, 4045 monomer bases of C, 4106 bases of G, and 5308 nucleotide bases of T are available as depicted in Figure 3.39. The monomer base count of T is lesser than the base count of A nucleotide. Also, the monomer bases of G nucleotide are higher compared to the monomer bases of C nucleotide in the considered gene sequence. Also, the dimers' distribution in a gene sequence with accession X92412.1 is depicted in the form of a bar illustration. In Figure 3.40, it is shown that dimers using T-T base pairs are more when compared with all other dimers in the gene sequence with accession X92412.1. There are maximum base counts with 2293 dimers of A-A base pairs in this gene sequence.

In this gene sequence with accession X92412.1, there are 1485 A-T dimers and 778 G-C dimers. G+C content is less compared to A+T dimers in this sequence, which indicates that it has a less number of genes. Figure 3.41 depicts the typical plot for monomer and dimer nucleotide densities for the nucleotide sequence with accession U01317.1. It is evident from the above plot that monomer densities for T nucleotide bases are present along the length of the sequence and its density is more than other nucleotide bases A, C, and G in the gene sequence with accession U01317.1.

Figure 3.42 shows a typical pie illustration for the monomer base counts distribution in a nucleotide sequence with accession U01317.1. A list of base nucleotide counts for the 5'–3' strand in a DNA sequence with accession U01317.1 is plotted to present nucleotide distribution. In this gene sequence, 22068 bases of A nucleotide, 14146 monomer bases of C, 14785 bases of G, and 22309 nucleotide bases of

T are available as depicted in Figure 3.42. The monomer base count of T is slightly lesser than the base count of A nucleotide. Also, the monomer bases of G nucleotide are higher compared to the monomer bases of C nucleotide in the considered gene sequence.

The dimers' distribution in a gene sequence with accession U01317.1 is also portrayed as a bar illustration. Figure 3.43 shows that dimers using T-T base pairs are more compared to all other dimers in the U01317.1 gene sequence.

There are maximum base counts with 7455 dimers of base pairs of A-A in this gene sequence.

There are 5950 A-T dimers and 2836 G-C dimers in this gene sequence with accession U01317.1. In this sequence, G+C content is less than A+T dimers, indicating that it has fewer genes. Figure 3.44 shows the typical plot for the nucleotide sequence with accession X51502.1 for monomer and dimer nucleotide densities. It is evident from the above plot that monomer densities for T nucleotide bases are present along the length of the sequence and its density is more than other nucleotide bases A, C, and G in the gene sequence with accession X51502.1. Figure 3.45 shows a typical pie illustration for the monomer base counts distribution in a nucleotide sequence with accession X51502.1.

The 5'–3' strand shows a list of the nucleotide base counts in a DNA sequence with accession X51502.1 as plotted below to depict the nucleotide distribution. In the gene sequence with accession X51502.1, 1245 bases of A nucleotide, 723 monomer bases of C, 754 bases of G, and 1069 nucleotide bases of T are available as depicted in Figure 3.45. The monomer base count for T is slightly lesser than the base count for A nucleotide. Also, the monomer bases of G nucleotide are higher compared to the monomer bases of C nucleotide in the considered gene sequence. Automatic cell identification using an invariant system in *Caenorhabditis elegans* sequence is discussed in Ref. [95].

Also, the dimers' distribution in a gene sequence with accession X51502.1 is depicted in the form of a bar illustration. In Figure 3.46, it is shown that dimers using T-T base pairs are more when compared with all other dimers in the gene sequence with accession X51502.1. There are maximum base counts with 435 dimers of A-A base pairs in this gene sequence. In this gene sequence with accession X51502.1, there are 314 A-T dimers and 144 G-C dimers. G+C content is less compared to A+T dimers in this sequence, which indicates that it has a less number of genes.

3.13.3 Performance Measures of Exon Prediction

Performance measures are carried out with the use of metrics like sensitivity (Sn), specificity (Sp), and precision (Pr) of LMS as well as its signed AEPs for Sn, Sp, and Pr calculation in the DNA sequence with accession AF009962. In order to identify the exon sections using DSP-based adaptive methods, few performance measures are definite based on changing the threshold level of the output spectrum. The amount of nucleotides situated as introns is identified as true negative (TN) in the exon identification phase and properly recognized as exons is termed as true positive (TP). Also, the total of exon segments as intron nucleotides is described as false negatives (FN), whereas these are actually measured as exon nucleotides as false positives (FP).

The quantity of exons found in exons is known as specificity (Sp), and the quantity of exons which are properly predicted as exons is measured as sensitivity (Sn). These parameters with their theory and expressions are given in Ref. [80]. Expressions such as specificity, sensitivity, and precision for performance measurements are therefore provided below.

a. *Sensitivity:* The fraction of exons correctly predicted over actual exons, and the ones actually existing in exon segments are determined by

$$S_n = \frac{\text{TP}}{\text{TP} + \text{FN}} = \frac{\text{No. of correct exons}}{\text{No. of actual exons}} \tag{3.62}$$

b. *Specificity:* The fraction of correct exons over the number of predicted exons is measured using

$$S_p = \frac{\text{TP}}{\text{TP} + \text{FP}} = \frac{\text{No. of correct exons}}{\text{No. of predicted exons}} \tag{3.63}$$

c. *Precision:* The fraction of correctly classified features to the total testing class of features determined using

$$P_r = \left[\frac{\text{TP} + \text{TN}}{\text{TP} + \text{FP} + \text{TN} + \text{FN}}\right] = \frac{\text{No. of correctly classified features}}{\text{Total testing class of features}} \tag{3.64}$$

where

TP = True Positive = amount which are correctly predicted as exons
TN = True Negative = amount which are correctly predicted as introns
FP = False Positive = amount of exons which are predicted as introns
FN = False Negative = amount of introns which are predicted as exons

3.13.4 Exon Prediction Results

Here, the performance measures of various AEPs developed using LMS-based variants are derived, analyzed, and compared. All the LMS-based algorithms include LMF, VSLMS, LMLS, and LLAD along with their signed variants to develop diverse AEPs. In contrast, an LMS-based AEP is also created.

To evaluate the output of various developed AEPs, we have considered the exon prediction results for ten gene sequences from the NCBI gene databank. Performances of sequence 5 with accession AF009962 for all sign-based LMS variants are measured using MATLAB as shown in Tables 3.9–3.13. The Sn, Sp, and Pr performances are evaluated at the 0.4–0.9 threshold values with a 0.05 interval. At 0.8 threshold value, the exon prediction appears better and performance measurement values using sign LMS-based techniques are shown in Table 3.9. From these values, it is evident that the performance measures of LMS-based AEP are just superior compared to SRLMS-based AEP with a less number of iterations.

For a larger number of iterations, the computational complexity of LMS is higher than SRLMS algorithm. In such cases, the SRLMS algorithm becomes efficient due

TABLE 3.9
Performance Measures of Sign LMS-Based AEPs for Gene Sequence with Accession AF009962

		Gene Sequence Serial Number									
Algorithm	Metric	1	2	3	4	5	6	7	8	9	10
LMS	Sn	0.6286	0.6384	0.6457	0.6273	0.6481	0.6162	0.6193	0.6241	0.6268	0.6202
	Sp	0.6435	0.6628	0.6587	0.6405	0.6518	0.6324	0.6529	0.6289	0.6452	0.5965
	Pr	0.5922	0.5894	0.5934	0.5858	0.5904	0.5786	0.5896	0.5856	0.5814	0.5761
SRLMS	Sn	0.5813	0.6023	0.6236	0.5473	0.5849	0.6072	0.5929	0.5915	0.5941	0.5848
	Sp	0.6261	0.6054	0.6084	0.6315	0.6105	0.6151	0.6145	0.6438	0.5650	0.5457
	Pr	0.5694	0.5727	0.5694	0.5586	0.5751	0.5686	0.5764	0.5753	0.5726	0.5525
SLMS	Sn	0.5313	0.5802	0.5937	0.5587	0.4762	0.5827	0.5364	0.5705	0.5340	0.5640
	Sp	0.5774	0.5745	0.5716	0.5756	0.5799	0.5633	0.5746	0.5726	0.5481	0.5243
	Pr	0.5334	0.5583	0.5670	0.5586	0.5710	0.5463	0.5345	0.5539	0.5362	0.5372
SSLMS	Sn	0.4413	0.4486	0.4835	0.4831	0.4514	0.4528	0.4927	0.4862	0.4562	0.4321
	Sp	0.5171	0.5171	0.5581	0.5257	0.5684	0.4765	0.5049	0.4686	0.4320	0.5152
	Pr	0.5278	0.5429	0.5587	0.5489	0.5704	0.5329	0.5132	0.5191	0.5052	0.5151

TABLE 3.10
Performance Measures of LMF and Its Sign-Based AEPs for Gene Sequence with Accession AF009962

		Gene Sequence Serial Number									
Algorithm	Metric	1	2	3	4	5	6	7	8	9	10
LMS	Sn	0.6286	0.6384	0.6457	0.6273	0.6481	0.6162	0.6193	0.6241	0.6268	0.6202
	Sp	0.6435	0.6628	0.6587	0.6405	0.6518	0.6324	0.6529	0.6289	0.6452	0.5965
	Pr	0.5922	0.5894	0.5934	0.5858	0.5904	0.5786	0.5896	0.5856	0.5814	0.5761
LMF	Sn	0.6694	0.6708	0.6686	0.6647	0.6705	0.6596	0.6546	0.6572	0.6682	0.6718
	Sp	0.6676	0.6791	0.6692	0.6635	0.6717	0.6632	0.6732	0.6594	0.6592	0.6592
	Pr	0.6412	0.6586	0.6518	0.6597	0.6512	0.6404	0.6514	0.6576	0.6535	0.6484
SRLMF	Sn	0.6513	0.6523	0.6594	0.6573	0.6649	0.6445	0.6432	0.6482	0.6548	0.6683
	Sp	0.6561	0.6743	0.6648	0.6568	0.6585	0.6475	0.6685	0.6516	0.6563	0.6486
	Pr	0.6394	0.6427	0.6494	0.6486	0.6451	0.6343	0.6431	0.6482	0.6465	0.6373
SLMF	Sn	0.6426	0.6492	0.6536	0.6431	0.6519	0.6327	0.6356	0.6325	0.6441	0.6523
	Sp	0.6487	0.6716	0.6614	0.6468	0.6485	0.6375	0.6602	0.6483	0.6518	0.6385
	Pr	0.6249	0.6372	0.6385	0.6386	0.6352	0.6252	0.6317	0.6337	0.6362	0.6293
SSLMF	Sn	0.6311	0.6422	0.6536	0.6375	0.6649	0.6241	0.6232	0.6412	0.6345	0.6457
	Sp	0.6361	0.6682	0.6597	0.6368	0.6385	0.6274	0.6567	0.6364	0.6496	0.6286
	Pr	0.6195	0.6227	0.6288	0.6286	0.6251	0.6143	0.6231	0.6283	0.6265	0.6182

TABLE 3.11
Performance Measures of VSLMS and Its Sign-Based AEPs for Gene Sequence with Accession AF009962

		Gene Sequence Serial Number									
Algorithm	Metric	1	2	3	4	5	6	7	8	9	10
LMS	Sn	0.6286	0.6384	0.6457	0.6273	0.6481	0.6162	0.6193	0.6241	0.6268	0.6202
	Sp	0.6435	0.6628	0.6587	0.6405	0.6518	0.6324	0.6529	0.6289	0.6452	0.5965
	Pr	0.5922	0.5894	0.5934	0.5858	0.5904	0.5786	0.5896	0.5856	0.5814	0.5761
VSLMS	Sn	0.6886	0.6797	0.6805	0.6796	0.6865	0.6796	0.6846	0.6872	0.6882	0.6818
	Sp	0.6792	0.6835	0.6887	0.6832	0.6787	0.6832	0.6789	0.6794	0.6795	0.6892
	Pr	0.6878	0.6793	0.6812	0.6814	0.6812	0.6804	0.6814	0.6876	0.6836	0.6884
SRVSLMS	Sn	0.6795	0.6723	0.6749	0.6682	0.6749	0.6672	0.6732	0.6782	0.6806	0.6692
	Sp	0.6684	0.6768	0.6685	0.6674	0.6625	0.6775	0.6642	0.6696	0.6676	0.6768
	Pr	0.6804	0.6686	0.6675	0.6643	0.6751	0.6694	0.6731	0.6747	0.6765	0.6793
SVSLMS	Sn	0.6679	0.6672	0.6674	0.6568	0.6641	0.6567	0.6673	0.6678	0.6756	0.6564
	Sp	0.6581	0.6576	0.6567	0.6474	0.6587	0.6632	0.6564	0.6569	0.6567	0.6653
	Pr	0.6692	0.6525	0.6423	0.6413	0.6654	0.6569	0.6673	0.6674	0.6676	0.6592
SSVSLMS	Sn	0.6567	0.6567	0.6572	0.6568	0.6641	0.6584	0.6607	0.6587	0.6644	0.6518
	Sp	0.6458	0.6455	0.6453	0.6417	0.6587	0.6503	0.6516	0.6504	0.6556	0.6549
	Pr	0.6369	0.6352	0.6346	0.6403	0.6654	0.6436	0.6463	0.6457	0.6475	0.6354

TABLE 3.12
Performance Measures of LMLS and Its Sign-Based AEPs for Gene Sequence with Accession AF009962

		Gene Sequence Serial Number									
Algorithm	Metric	1	2	3	4	5	6	7	8	9	10
LMS	Sn	0.6286	0.6384	0.6457	0.6273	0.6481	0.6162	0.6193	0.6241	0.6268	0.6202
	Sp	0.6435	0.6628	0.6587	0.6405	0.6518	0.6324	0.6529	0.6289	0.6452	0.5965
	Pr	0.5922	0.5894	0.5934	0.5858	0.5904	0.5786	0.5896	0.5856	0.5814	0.5761
LMLS	Sn	0.7702	0.7691	0.7692	0.7638	0.7679	0.7725	0.7774	0.7787	0.7791	0.7792
	Sp	0.7698	0.7585	0.7582	0.7791	0.7797	0.7763	0.7738	0.7764	0.7685	0.7682
	Pr	0.7753	0.7463	0.7764	0.7685	0.7658	0.7792	0.7786	0.7767	0.7763	0.7764
SRLMLS	Sn	0.7615	0.7656	0.7617	0.7557	0.7565	0.7553	0.7676	0.7452	0.7556	0.7587
	Sp	0.7545	0.7423	0.7323	0.7556	0.7692	0.7642	0.7557	0.7512	0.7543	0.7523
	Pr	0.7675	0.7342	0.7632	0.7543	0.7554	0.7587	0.7574	0.7553	0.7642	0.7652
SLMLS	Sn	0.7563	0.7565	0.7561	0.7505	0.7516	0.7525	0.7567	0.7345	0.7515	0.7478
	Sp	0.7514	0.7522	0.7292	0.7434	0.7569	0.7464	0.7475	0.7501	0.7454	0.7385
	Pr	0.7567	0.7431	0.7463	0.7357	0.7485	0.7458	0.7417	0.7455	0.7567	0.7452
SSLMLS	Sn	0.7456	0.7506	0.7456	0.7450	0.7451	0.7452	0.7516	0.7314	0.7433	0.7347
	Sp	0.7451	0.7472	0.7219	0.7414	0.7526	0.7415	0.7397	0.7450	0.7415	0.7318
	Pr	0.7434	0.7383	0.7346	0.7285	0.7348	0.7374	0.7341	0.7385	0.7446	0.7345

TABLE 3.13
Performance Measures of LLAD and Its Sign-Based AEPs for Gene Sequence with Accession AF009962

		Gene Sequence Serial Number									
Algorithm	Metric	1	2	3	4	5	6	7	8	9	10
LMS	Sn	0.6286	0.6384	0.6457	0.6273	0.6481	0.6162	0.6193	0.6241	0.6268	0.6202
	Sp	0.6435	0.6628	0.6587	0.6405	0.6518	0.6324	0.6529	0.6289	0.6452	0.5965
	Pr	0.5922	0.5894	0.5934	0.5858	0.5904	0.5786	0.5896	0.5856	0.5814	0.5761
LLAD	Sn	0.8163	0.8188	0.8165	0.8294	0.8294	0.8126	0.8208	0.8172	0.8279	0.8284
	Sp	0.8223	0.8265	0.8252	0.8262	0.8165	0.8234	0.8215	0.8248	0.8218	0.8172
	Pr	0.8145	0.8202	0.8147	0.8278	0.8242	0.8175	0.8198	0.8152	0.8263	0.8256
SRLLAD	Sn	0.8092	0.8102	0.8087	0.8154	0.8162	0.8078	0.8173	0.8095	0.8206	0.8211
	Sp	0.8184	0.8178	0.8195	0.8208	0.8086	0.8185	0.8182	0.8187	0.8167	0.8125
	Pr	0.8086	0.8164	0.8112	0.8212	0.8185	0.8083	0.8116	0.8074	0.8178	0.8196
SLLAD	Sn	0.7914	0.8010	0.7982	0.8115	0.8016	0.8007	0.8017	0.8009	0.8120	0.8021
	Sp	0.8025	0.8017	0.8019	0.8120	0.7968	0.8018	0.8018	0.8118	0.8017	0.8052
	Pr	0.7968	0.8106	0.8001	0.8021	0.8018	0.7948	0.7961	0.8007	0.7921	0.8003
SSLLAD	Sn	0.7891	0.7952	0.7854	0.8011	0.7952	0.7872	0.7982	0.7945	0.8012	0.7902
	Sp	0.7954	0.7971	0.7925	0.8008	0.7896	0.7895	0.7881	0.8011	0.7931	0.7985
	Pr	0.7886	0.8012	0.7946	0.7937	0.7965	0.7894	0.7906	0.7884	0.7843	0.7916

to low complexity in performing the computations and in terms of exon locating ability. The signum function present in all signed versions of LMS reduces the computational complexity, and thus, all signed versions predict the exon locations more accurately.

Of all these algorithms, SRLMS-based AEP is effective in terms of accurate exon prediction when compared to LMS and its other signed variants with sensitivity Sn 0.5849 (58.49%), specificity Sp 0.6105 (61.05%), and precision Pr 0.5751 (57.51%). The exon prediction seems better at threshold 0.8 for sign regressor–based AEP for a gene sequence with accession AF009962. A typical PSD plot for LMS and its sign regressor–based variant is shown in Figure 3.47. It demonstrates the expected exon sites of sequence 5 with LMS-based adaptive algorithms. From these plots, it is evident that coding regions are not correctly predicted by LMS-based AEP. This algorithm creates some location prediction ambiguities in the form of short peaks for exon location and some unwanted peaks at different positions in the power spectrum. The PSD plots for LMS and its signed variants are shown in Figure 3.47a–d.

Figure 3.47a identifies some undesirable peaks at various locations using LMS algorithm. The actual exon position 3934-4581 is not simultaneously predicted accurately using LMS algorithm. It is clear that SRLMS-based AEP predicts the exon location with a sharp peak accurately compared to all other LMS variants. Using LMS-based adaptive algorithms like SRLMS, SLMS, and SSLMS, prediction measure values for sensitivity, specificity, and precision from Table 3.9 are observed a bit inferior than the LMS adaptive algorithm, which are much better in the case of normalized algorithms.

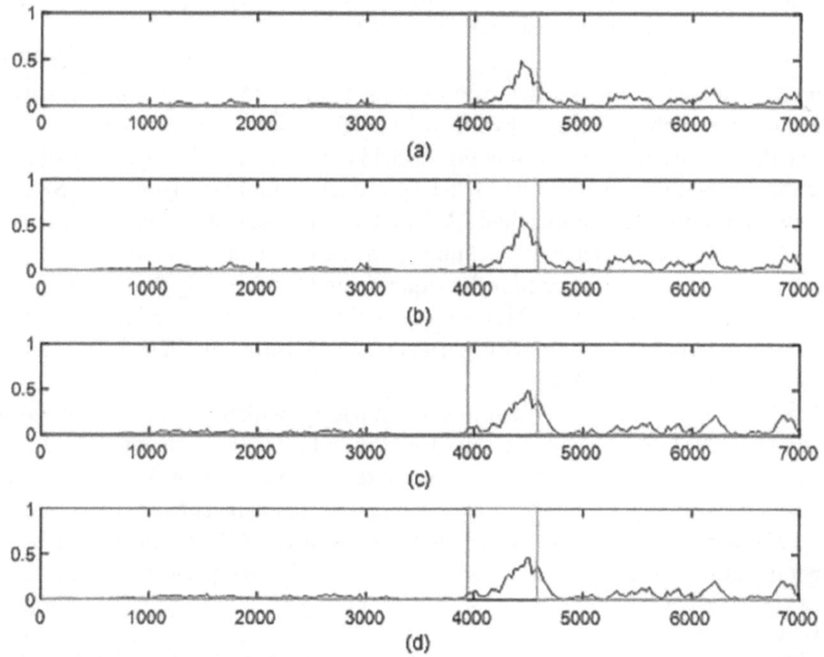

FIGURE 3.47 PSD plots with the location of exon (3934-4581) for a DNA sequence with accession AF009962 predicted using (a) LMS-based AEP, (b) SRLMS-based AEP, (c) SLMS-based AEP, and (d) SSLMS-based AEP. (*Relative base location is taken on x-axis, and power spectrum is taken on y-axis.*)

Due to use of signum function in SRLMS-based AEP, it offers good exon location ability and low computational complexity compared to LMS and other signed variants. The performance measures of sequence 5 with accession AF009962 for all sign-based LMF variants are measured using MATLAB as shown in Table 3.10. The Sn, Sp, and Pr parameters measure the efficiency at 0.4–0.9 threshold values with an interval of 0.05. At a threshold value of 0.8, the exon prediction appears to be better and output values using signed LMF-based algorithms are as shown in Table 3.10. From the values, it is evident that the performance measures of LMF-based AEP are just inferior compared to SRLMF-based AEP with a less number of iterations. For a larger number of iterations, the computational complexity of LMF is higher than SRLMF algorithm.

In such cases, the SRLMF algorithm becomes efficient due to low complexity in performing the computations and in terms of exon locating ability. The signum function present in all signed versions of LMF reduces the computational complexity, and thus, all signed versions predict the exon locations more accurately. Of all these algorithms, SRLMF-based AEP is effective in terms of accurate exon prediction when compared to LMS, LMF, and its other signed variants with sensitivity Sn 0.6649 (66.49%), specificity Sp 0.6585 (65.85%), and precision Pr 0.6451 (64.51%).

The exon prediction seems to be better at threshold 0.8 for sign regressor LMF–based AEP for a gene sequence with accession AF009962.

A typical PSD plot for LMF and its sign-based variants is shown in Figure 3.48. In the case of LMF versions, the exon location at 3934-4581 with high intensity in PSD plot is precisely anticipated by the SRLMF algorithm. These PSD plots of LMF and the signed variants are shown in Figure 3.48a–d. The tracking capability of these algorithms is better than LMS due to the sign function in these algorithms. SRLMF has better convergence features than LMS and its other signed variants.

For SSLMF, the efficiency of locating exon sections is lower compared to other signed variants due to the clipped input sequence and feedback signal. Therefore, it is evident that SRLMF-based AEP is a better candidate in exon prediction applications based on computational complexity, performance measures in Table 3.10, and exon prediction plots in Figure 3.48.

Similarly, the performance measures of sequence 5 with accession AF009962 for all sign-based VSLMS variants measured using MATLAB are as shown in Table 3.11. The values obtained for Sn, Sp, and Pr measures are evaluated at 0.4–0.9 threshold values with an interval of 0.05. The exon forecast seems to be better at threshold 0.8, and performance measurement values using sign-based VSLMS algorithms are as shown in Table 3.11. From these values, it is evident that the performance measures

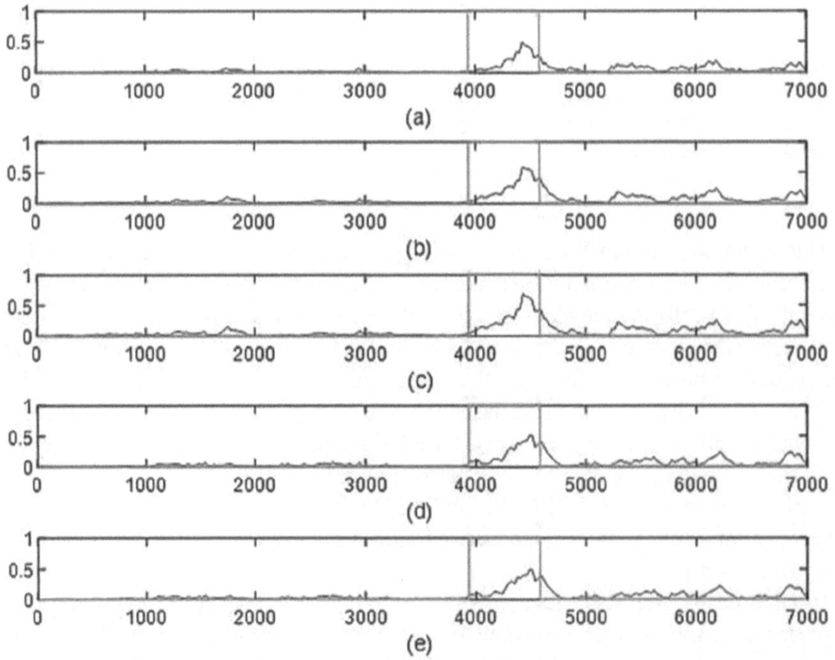

FIGURE 3.48 PSD plots with the location of exon (3934-4581) for a DNA sequence with accession AF009962 predicted using (a) LMS-based AEP, (b) LMF-based AEP, (c) SRLMF-based AEP, (d) SLMF-based AEP, and (e) SSLMF-based AEP. (*Relative base location is taken on x-axis, and power spectrum is taken on y-axis.*)

of VSLMS-based AEP are just inferior compared to SRVSLMS-based AEP with a less number of iterations. For a larger number of iterations, the computational complexity of VSLMS is higher than SRVSLMS algorithm.

In such cases, SRVSLMS algorithm becomes efficient due to low complexity in performing the computations and in terms of exon locating ability. The signum function present in all signed versions of VSLMS reduces the computational complexity, and thus, all signed versions predict the exon locations more accurately.

Of all these algorithms, SRVSLMS-based AEP is effective in terms of accurate exon prediction when compared to LMS, VSLMS, and its other signed variants with sensitivity Sn 0.6749 (67.49%), specificity Sp 0.6625 (66.25%), and precision Pr 0.6751 (67.51%). At a threshold value of 0.8, the exon prediction appears to be better for sign regressor VSLMS–based AEP for a gene sequence with accession AF009962. In the case of VSLMS versions, SRVSLMS algorithm located the exon position precisely at 3934-4581 with great intensity and a sharp peak in PSD plot is observed. These PSD plots for VSLMS and its signed variants are shown in Figure 3.49a–d.

The ability of tracking for these algorithms is better than LMS algorithm due to the sign function present in it. Also, the convergence performance for SRVSLMS is better than LMS and other signed variants. With SSVSLMS, the efficiency of exon prediction is lower than its other signed variants owing to the clipped input sequence

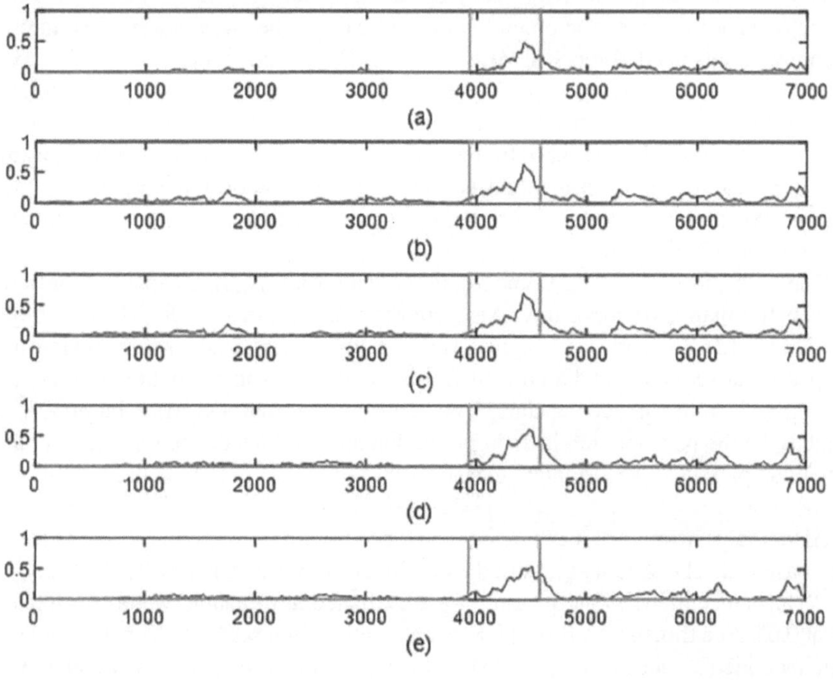

FIGURE 3.49 PSD plots with the location of exon (3934-4581) for a DNA sequence with accession AF009962 predicted using (a) LMS-based AEP, (b) VSLMS-based AEP, (c) SRVSLMS-based AEP, (d) SVSLMS-based AEP, and (e) SSVSLMS-based AEP. (*Relative base location is taken on x-axis, and power spectrum is taken on y-axis.*)

and feedback signal. Thus, it is pragmatic that the SRVSLMS-based AEP is the best candidate in exon prediction applications, based on computational complexity, performance measures in Table 3.6, and exon prediction plots in Figure 3.49.

Likewise, the performance measures of sequence 5 with accession AF009962 for all sign-based LMLS variants measured using MATLAB are shown in Table 3.12.

The performance values of Sn, Sp, and Pr appear to be measured at 0.4–0.9 threshold values at a rate of 0.05. The exon prediction at a threshold level of 0.8 appears better, and values of performance metrics using sign LMLS–based algorithms are better as shown in Table 3.12. From the values, it is evident that the performance measures of LMLS-based AEPs are just inferior compared to SRLMLS-based AEP with a less number of iterations. For a larger number of iterations, the computational complexity of LMLS is higher than SRLMLS algorithm.

In such cases, SRLMLS algorithm becomes efficient due to low complexity in performing the computations and in terms of exon locating ability. The signum function present in all signed versions of LMLS reduces the computational complexity, and thus, all signed versions predict the exon locations more accurately. Of all these algorithms, SRLMLS-based AEP is effective in terms of accurate exon prediction when compared to LMS, LMLS, and its other signed variants with sensitivity Sn 0.7565 (75.65%), specificity Sp 0.7692 (76.92%), and precision Pr 0.7554 (75.54%).

These values are only inferior than those obtained using LMLS-based AEP. But it is a better candidate for exon precipitation applications, because of a smaller amount of computations needed for SRLMLS-based AEP. At a threshold value of 0.8, exon forecast using sign regressor LMLS–based AEP for a gene sequence with accession AF009962 appears to be better. A typical PSD plot for LMLS and its sign-based variants is shown in Figure 3.50. In the case of LMLS versions, SRLMLS algorithm exactly predicted the exon location at 3934-4581 with good intensity and a sharp peak in PSD plot is observed. These PSD plots for LMLS and its signed variants are shown in Figure 3.50a–d.

Due to the presence of sign function in these algorithms, their capability of tracking is better than LMS algorithm. Also, convergence features of SRLMLS are better than LMS, LMLS, and other signed variants. In SSLMLS algorithm, owing to a clipped input sequence and a clipped feedback signal, exon prediction efficiency is lower than its other signed variants. It is therefore clear that SRLMLS-based AEP is found to be the best solution in exon prediction applications depending on computational complexities, performance measures in Table 3.12, and exon prediction plots in Figure 3.50.

Also, the performance measures of sequence 5 with accession AF009962 for all sign-based LLAD variants measured using MATLAB are shown in Table 3.13.

Performance metrics Sn, Sp, and Pr are evaluated at threshold values from 0.4 to 0.9 at 0.05. At a threshold value of 0.8, exon identification seems to be better and performance metric values using LLAD-based algorithms are as shown in Table 3.12. From these values, it is evident that the performance measures of LLAD-based AEPs are just inferior compared to SRLLAD-based AEP with a less number of iterations. For a larger number of iterations, the computational complexity of LLAD is higher than SRLLAD algorithm.

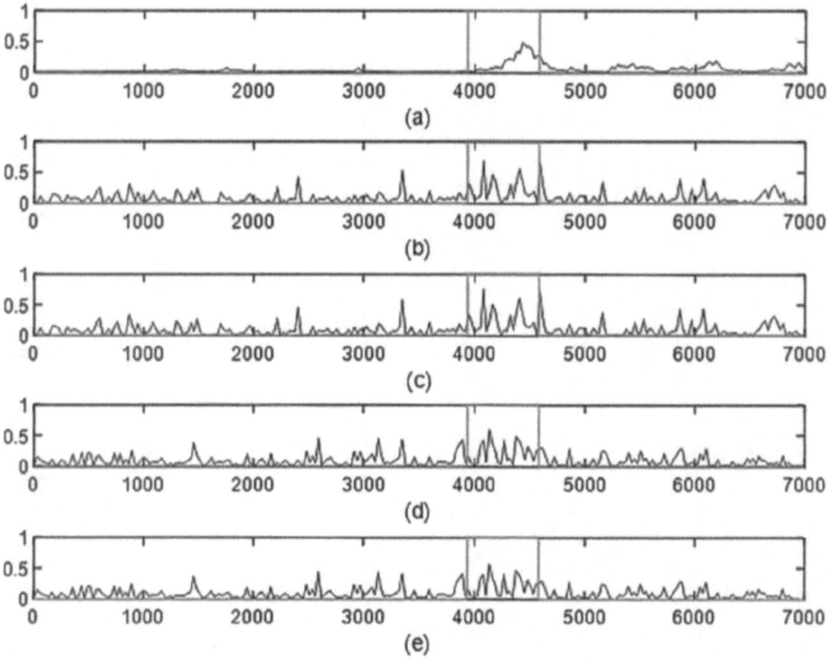

FIGURE 3.50 PSD plots with the location of exon (3934-4581) for a DNA sequence with accession AF009962 predicted using (a) LMS-based AEP, (b) LMLS-based AEP, (c) SRLMLS-based AEP, (d) SLMLS-based AEP, and (e) SSLMLS-based AEP. (*Relative base location is taken on x-axis, and power spectrum is taken on y-axis.*)

In such cases, the SRLLAD algorithm becomes efficient due to low complexity in performing the computations and in terms of exon locating ability. The signum function present in all signed versions of LLAD reduces the computational complexity, and thus, all signed versions predict the exon locations more accurately.

Of all these algorithms, SRLLAD-based AEP is effective in terms of accurate exon prediction when compared to LMS, LLAD, and its other signed variants with sensitivity Sn 0.8162 (81.62%), specificity Sp 0.8086 (80.86%), and precision Pr 0.8185 (81.85%). These values are just lower than those obtained using LLAD-based AEP. But owing to fewer computations needed for SRLLAD-based AEP, it is chosen to be a better candidate for exon prediction applications. At a threshold value of 0.8, exon identification appears to be better for SRLLAD-based AEP for a gene sequence with accession AF009962.

A typical PSD plot for LLAD and its sign-based variants is shown in Figure 3.51. In the case of LMLS versions, SRLLAD algorithm located exon position with a greater accuracy at 3934-4581, and good intensity and a sharp peak in PSD plot are observed. These PSD plots for LLAD and its signed variants are shown in Figure 3.51a–d. Due to the sign function present in these algorithms, exon tracking is better than LMS algorithm. SRLLAD also has better convergence features than LMS, LLAD, and its other signed variants.

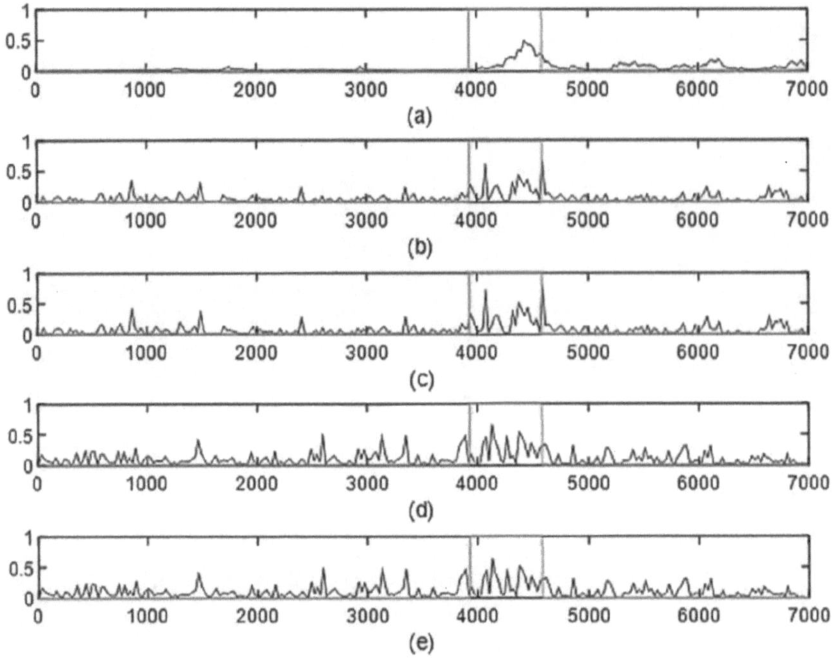

FIGURE 3.51 PSD plots with the location of exon (3934-4581) for a DNA sequence with accession AF009962 predicted using (a) LMS-based AEP, (b) LLAD-based AEP, (c) SRLLAD-based AEP, (d) SLLAD-based AEP, and (e) SSLLAD-based AEP. (*Relative base location is taken on x-axis, and power spectrum is taken on y-axis.*)

In the case of SSLLAD, the prediction efficiency of exon position is inferior compared to its other signed variants owing to the clipped input sequence and clipped feedback signal. It has been shown that SRLLAD-based AEP is the leading candidate for exon prediction applications relative to current LMS and any other signed LMS-based adaptive algorithms based on computational complexity, the performance measures in Table 3.8, and the exon prediction plots in Figure 3.51.

3.14 CONCLUSIONS

Predicting the accurate exon segments in healthcare has several applications and more significance. In this chapter, we have developed LMS-based weight update variants of adaptive algorithms. These are used to develop various AEPs for accurate prediction of exon locations in real gene sequences of *Homo sapiens*. A novel adaptive methodology for exon identification is suggested here. Varieties of AEPs are examined with five LMS-based algorithms and their sign-based variants on original DNA datasets of *Homo sapiens* taken from the NCBI gene databank.

Here, theoretical considerations of adaptive filtering techniques for DNA analysis, introduction to adaptive filtering along with properties of adaptive algorithms, need for development of AEPs, and structure of AEP used for DNA analysis are discussed.

Also, the familiar LMS algorithm is discussed elaborately in this chapter. This algorithm is considered as the reference algorithm, and various proposed realizations with this algorithm are presented. We inspect the possibility to enhance the existing methods in order to achieve good accuracy in exon prediction, better convergence performance, and low computational complexity.

To achieve this, reduction of computational complexity and improvement in exon tracking ability are desirable. This can be achieved by either clipping input data or estimation error. Three signed algorithms are derived based on this clipping and utilized here. Now, we have extended the LMS-based adaptive algorithms to three versions of sign-based adaptive algorithms. This is the first category of the adaptive algorithms we have considered. Totally, twenty AEPs are developed and tested on ten real gene sequences of *Homo sapiens* taken from the NCBI databank. These include LMS, LMF, VSLMS, LMLS, and LLAD algorithms along with their fifteen sign-based algorithms, i.e., SRLMS, SLMS, SSLMS, SRLMF, SLMF, SSLMF, SRVSLMS, SVSLMS, SSVSLMS, SRLMLS, SLMLS, SSLMLS, SRLLAD, SLLAD, and SSLLAD algorithms.

Plots for convergence characteristics are provided, and PSD and their computational complexities and measures like sensitivity (Sn), specificity (Sp), and precision (Pr) are also calculated to assess the performance of different AEPs. Among the five LMS-based algorithms, LLAD algorithm performs better compared to all other algorithms in terms of exon tracking ability and low computational complexity. Among the sign-based LMS variants, all their sign regressor versions performed better. All their SA- and SSA-based AEPs are slightly inferior to their LMS counterparts. Overall, all the proposed sign LMS–based AEPs deliver greater performance than the existing LMS pertaining to computational complexity and metrics like sensitivity, precision, and specificity attained using a gene sequence with accession AF009962 at a threshold value of 0.8. Based on performance measures, exon prediction accuracy, computational complexities, and convergence efficiency of different LMS-based AEPs, it is evident that SRLLAD-based AEP is more effective in exon prediction applications among all LMS-based AEPs.

4 Normalization-Based Realization of Adaptive Filtering Techniques for Exon Prediction

4.1 INTRODUCTION

Exactly identifying the exon gene sequence segments depending upon the Three Base Periodicity (TBP) property is an essential task in genomics for the diagnosis of illnesses and the design of medicines. In order to achieve fast convergence and good filtering capability, we introduce normalization-based AEPs and their sign-based versions to reduce computational burden. The familiar LMS technique has wide applications due to its simplicity and robustness. To choose step size parameter, the LMS filter requires prior understanding of input power level for stabilization and convergence. Since the level of power input remains generally one of the statistically unknowns, this remains generally predictable from information before the starting process of adaptation. But it suffers with two primary drawbacks in practical circumstances. These can be well explained from its weight recursion.

From the LMS weight update relation provided in (3.17), it is evident that the data input vector is directly proportional to the weight updating process. Other is fixing size of the step. Practically, design of an algorithm must be in a way that powerful and weak signals must be addressed. Therefore, the coefficients of tap should be adapted consequently based on input and output changes of filter. The LMS algorithm therefore suffers from a setback of gradient noise amplification when the data input vector is big. It is possible to use normalized LMS (NLMS) algorithm to prevent this issue. In each iteration, adjustment applied to the coefficient of filter weight vector is normalized with respect to Euclidian input vector's squared norm. Because of the fixed size of step, the speed of convergence is low in conventional LMS technique. In order to improve the convergence performance of LMS filter, a variable convergence factor is the best solution without the use of estimates of the input signal correlation matrix [97].

Size of the step alters iteratively due to normalization, and the instantaneous value of input data vector coefficients is proportional to the opposite of the total estimated energy. Generally, NLMS converges faster than LMS because it uses a varying convergence factor that minimizes an instantaneous error at output [98–101]. Several contributions on NLMS are given in Refs. [102–105]. Other strategies for varying the step size are reported in Refs. [106–110]. The summary of algorithms used in this chapter is given in Table 4.1.

TABLE 4.1
List of Algorithms Used in Chapter 4

S. No.	Acronym	Name	Advantage
1	NLMS	Normalized least mean square	Fast convergence, avoids gradient noise amplification problem
2	NSRLMS	Normalized sign regressor least mean square	Fast convergence, avoids gradient noise amplification problem, only one multiplication needed to compute the recursion
3	NSLMS	Normalized sign least mean square	Fast convergence, avoids gradient noise amplification problem, it requires L MACs
4	NSSLMS	Normalized sign-sign least mean square	Fast convergence, avoids gradient noise amplification problem, zero multiplications required, enter recursion is computed with addition with sign check operation
5	MNLMS	Maximum normalized least mean square	Low computational complexity, faster convergence than LMS, avoids gradient noise amplification problem
6	MNSRLMS	Maximum normalized sign regressor least mean square	Significantly low computational complexity, avoids gradient noise amplification problem, better stability
7	MNSLMS	Maximum normalized sign least mean square	Low computational complexity, avoids gradient noise amplification problem, low steady-state error and better convergence
8	MNSSLMS	Maximum normalized sign-sign least mean square	Fast convergence, avoids gradient noise amplification problem, less computations
9	ENLMS	Error-normalized least mean square	Fast convergence, minimizes signal distortion
10	ENSRLMS	Error-normalized sign regressor least mean square	Fast convergence, minimizes signal distortion, only one multiplication needed to compute the recursion
11	ENSLMS	Error-normalized sign least mean square	Fast convergence, minimizes signal distortion, it requires L MACs
12	ENSSLMS	Error-normalized sign-sign least mean square	Fast convergence, minimizes signal distortion, zero multiplications required, enter recursion is computed with addition with sign check operation
13	MENLMS	Maximum error-normalized least mean square	Low computational complexity, faster convergence than LMS, avoids gradient noise amplification problem
14	MENSRLMS	Maximum error-normalized sign regressor least mean square	Significantly low computational complexity, avoids gradient noise amplification problem, better stability
15	MENSLMS	Maximum error-normalized sign least mean square	Low computational complexity, avoids gradient noise amplification problem, low steady-state error and convergence better than LMS
16	MENSSLMS	Maximum error-normalized sign-sign least mean square	Fast convergence, avoids gradient noise amplification problem, less computations

(Continued)

TABLE 4.1 (*Continued*)
List of Algorithms Used in Chapter 4

S. No.	Acronym	Name	Advantage
17	NLMF	Normalized least mean fourth	Fast convergence, avoids gradient noise amplification problem
18	NSRLMF	Normalized sign regressor least mean fourth	Fast convergence, avoids gradient noise amplification problem, only one multiplication needed to compute the recursion
19	NSLMF	Normalized sign least mean fourth	Fast convergence, avoids gradient noise amplification problem, it requires L MACs
20	NSSLMF	Normalized sign-sign least mean fourth	Fast convergence, avoids gradient noise amplification problem, zero multiplications required, enter recursion is computed with addition with sign check operation
21	MNLMF	Normalized least mean fourth	Low computational complexity, faster convergence than LMS, avoids gradient noise amplification problem
22	MNSRLMF	Maximum normalized sign regressor least mean fourth	Significantly low computational complexity, avoids gradient noise amplification problem, better stability
23	MNSLMF	Maximum normalized sign least mean fourth	Low computational complexity, avoids gradient noise amplification problem, low steady-state error and convergence better than LMS
24	MNSSLMF	Maximum normalized sign-sign least mean fourth	Fast convergence, avoids gradient noise amplification problem, less computations
25	VNLMS	Variable step size normalized least mean square	Fast convergence, avoids gradient noise amplification problem, variable step size
26	VNSRLMS	Variable step size normalized sign regressor least mean square	Fast convergence, avoids gradient noise amplification problem, only one multiplication needed to compute the recursion, variable step size
27	VNSLMS	Variable step size normalized sign least mean square	Fast convergence, avoids gradient noise amplification problem, it requires L MACs, variable step size
28	VNSSLMS	Variable step size normalized sign-sign least mean square	Fast convergence, avoids gradient noise amplification problem, zero multiplications required, variable step size
29	MVNLMS	Maximum variable step size normalized least mean square	Low computational complexity, faster convergence than LMS, avoids gradient noise amplification problem
30	MVNSRLMS	Maximum variable step size normalized sign regressor least mean square	Significantly low computational complexity, avoids gradient noise amplification problem, better stability
31	MVNSLMS	Maximum variable step size normalized sign least mean square	Low computational complexity, avoids gradient noise amplification problem, low steady-state error and convergence better than LMS
32	MVNSSLMS	Maximum variable step size normalized sign-sign least mean square	Fast convergence, avoids gradient noise amplification problem, less computations

4.2 NORMALIZED ADAPTIVE ALGORITHMS

The normalized version of LMS overcomes the two drawbacks discussed in the above section. Consider the level of signal variation at the output of filter by choosing the size of normalized step parameter which results in a stable and quick converging algorithm. In the normalized variant of LMS algorithm, a squared norm equivalent factor normalizes the correlation between error and input reference [111]. Also, we have considered normalized variants of LMF and VSLMS, which results in normalized LMF (NLMF) and variable step size NLMS (VNLMS), respectively. To minimize computational difficulty, and improve the convergence performance and exon tracking ability, we have extended all these normalized variants and considered their signed versions in this chapter.

4.3 NORMALIZED LMS (NLMS) ALGORITHM

The NLMS algorithm is another class of adaptive algorithms used to train adaptive filter coefficients. The choice of step size is one of the problems with LMS adaptive filter design and implementation. The NLMS algorithm is regarded as a unique LMS algorithm application that takes into consideration the signal-level change at filter input and select a normalized step size parameter that will make it stable and rapidly converging adaptive algorithm [112]. The mathematical modeling of NLMS algorithm is presented in Table 4.2.

w(*n*) is the tap-weight vector with the taps stored in a row vector

$$=\left[w(n)\ w(n-1)\ \dots\ w(n-T+1)\right]^T$$

The expression for estimation error *e*(*n*) is given by

$$e(n)=d(n)-\ w^T(n).\ x(n)$$

The relation for mean square error or cost function *J*(*n*) is

$$J(n)=E\left[\left|d(n)-y(n)\right|^2\right]=E\left[\left|e(n)\right|^2\right]$$

Therefore, *e**(*n*) is obtained as

$$e^*(n)=\left(1-2\ \mu(n)\ x^T(n)\ x(n)\ e(n)\right) \tag{4.3}$$

where $\mu(n)$ is a normalized step size with $0<\mu'<2$. Minimizing $(e^*(n))^2$ with respect to $\mu(n)$ results in the following expression:

$$\mu(n)=\frac{1}{2\ x^T(n)\ x(n)} \tag{4.4}$$

which forces *e**(*n*) to zero. The LMS algorithm for stationary processes converges in a mean, if $0<\mu<\frac{2}{\lambda_{\max}}$, and converges in the mean square, if $0<\mu<\frac{2}{tr(R_x)}$;

TABLE 4.2
Mathematical Modeling of NLMS Algorithm

Parameters: T = number of taps (i.e. filter length)

μ = step size parameter

Let the tap input be $x(n)$ and filter length T is moderate to large.

Initialization: Set $w(0) = 0$ as the initial condition.

Data: Given $x(n) = T$-by-1 tap input vector to filter n_2 at time n

$$= \left[x(n), x(n-1)\ldots\ x(n-T+1)\right]^T$$

$d(n)$ is desired response at time n, μ' is the adaptation constant, error signal is defined as $e(n)$, ε is a small positive parameter to avoid denominator become too small, and $(.)^T$ is the transpose of (.).

To be computed:

$w(n+1)$ = estimate of tap-weight vector at time $n+1$

Computation:

The LMS recursion expression is given by

$$w(n+1) = w(n) + 2\mu(n)\ x(n)e(n) \tag{4.1}$$

where $\mu(n)$ is the time variable step size parameter, chosen to minimize the extent of a posteriori error given by

$$e^*(n) = d(n) - w^T(n+1)x(n) \tag{4.2}$$

is minimized in magnitude.

From (4.4) in (4.1), the resultant expression is obtained as

$$w(n+1) = w(n) + \frac{1}{2x^T(n)x(n)}e(n)x(n) \tag{4.5}$$

Setting the time varying size of step parameter as

$$\mu(n) = \frac{\mu'}{\boldsymbol{x}^T(n)\boldsymbol{x}(n)} = \frac{\mu'}{\left\|\boldsymbol{x}(n)\right\|^2} \tag{4.7}$$

A small positive constant parameter ε is added to prevent denominator from becoming too low, and hence, (4.7) becomes

$$\mu(n) = \frac{\mu'}{\varepsilon + \left\|\boldsymbol{x}(n)\right\|^2} \tag{4.8}$$

By replacing μ in the LMS weight vector update equation with $\mu(n)$ from (4.8), the weight update relation for NLMS is written as

$$\boldsymbol{w}(n+1) = \boldsymbol{w}(n) + \frac{\mu'}{\varepsilon + \left\|\boldsymbol{x}(n)\right\|^2}e(n)\boldsymbol{x}(n) \tag{4.10}$$

Thus, the weight update relation of maximum NLMS for $x_{\text{Li}} \neq 0$ and $\varepsilon \neq 0$ becomes

$$\boldsymbol{w}(n+1) = \boldsymbol{w}(n) + \frac{\mu'}{\varepsilon + \max\left(\left\|\boldsymbol{x}(n)\right\|\right)^2}e(n)\boldsymbol{x}(n) \tag{4.13}$$

however, because R_x is usually unknown to be either, λ_{max} or R_x, the bounds need to be estimated to apply it.

The step size bound for mean square LMS convergence is

$$0 < \mu < \frac{2}{x^T(n)x(n)} \tag{4.6}$$

The adaptation constant μ' is dimensionless for the NLMS filter, whereas the LMS filter adaptation constant μ has inverse power dimensions. The NLMS filter can be viewed as an LMS filter with a parameter of time-varying step size. NLMS introduces its own issue to resolve the gradient noise amplification issue associated with LMS filter [113].

Numerical problems can occur here, and we have to split the squared norm $\|x(n)\|^2$. To solve this issue, a small positive parameter constant ε is introduced to avoid denominators that are too small and a step parameter that is too high. Replacing μ in the LMS weight relation using $\mu(n)$ from (4.7) leads to NLMS expression:

$$w(n+1) = w(n) + \frac{\mu'}{\|x(n)\|^2} e(n)\, x(n) \tag{4.9}$$

The noise amplification problem is reduced thru normalization of LMS step size by $\|x(n)\|^2$ in NLMS. The flow chart for maximum NLMS algorithm is as depicted in Figure 4.1. Comparing (3.17) and (4.10), a scaled variant of LMS algorithm is weight update relation for NLMS. Hence, variation in the size of weight vector $w(n)$ remains inversely proportional to that of the $x(n)$ information vector norm. Smaller step size values result from this normalization compared to the standard LMS. NLMS usually exhibits faster convergence compared to LMS, as its factor of convergence is used to minimize instantaneous error at output [123].

So as to further diminish difficulty in performing computations, normalized block processing techniques are considered in this chapter. But at this time, we consider overlapping blocks. The additional computations required to computing the variable step size for all the algorithms discussed in the previous sections can be further reduced by calculating a variable step parameter using partitioning of input data to blocks and maximum magnitude of every block. This method is also termed as maximum normalization, and the resulting algorithm is the maximum NLMS (MNLMS) algorithm.

Now, the MNLMS algorithm with its weight update relation for $x_{Li} \neq 0$ and $\varepsilon = 0$ is written from (4.7) as

$$w(n+1) = w(n) + \frac{\mu'}{x_{Li}^2} x(n)\, e(n) \tag{4.11}$$

Here, we have also considered the weight update relation for the sign regressor form of maximum SRNLMS algorithm which is written from (4.7) as

$$w(n+1) = w(n) + \frac{\mu'}{x_{Li}^2} \text{Sign}\{x(n)\}\, e(n) \tag{4.12}$$

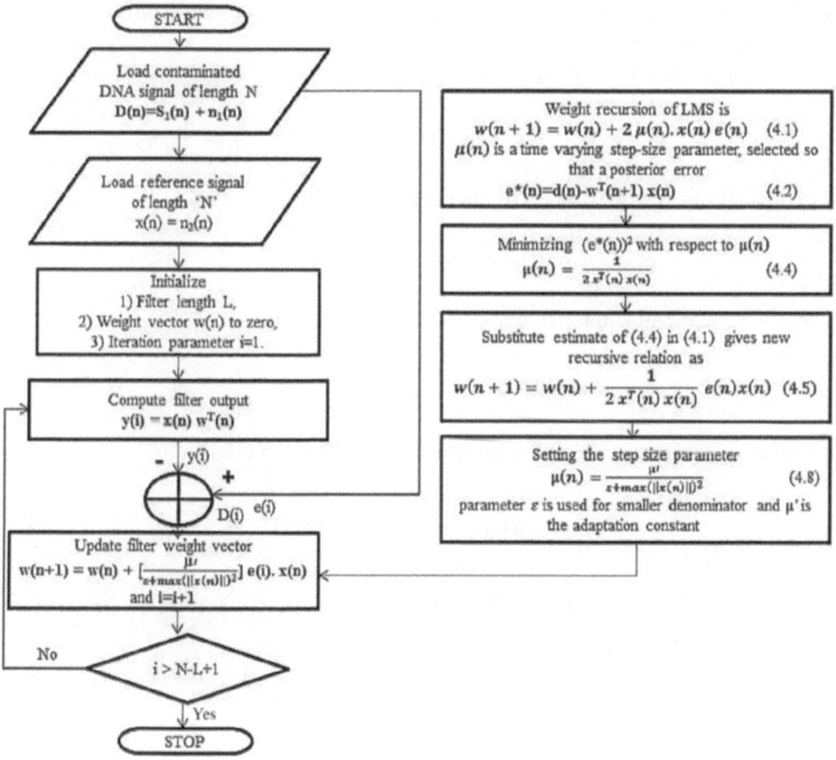

FIGURE 4.1 Flow chart of AEP based on maximum NLMS algorithm.

where $x_{Li} = \max\{|x_k|, k \in Z_i'\}, Z_i' = \{iL,\ iL+1, \ldots, iL+L-1\}, i \in Z,$ and for $x_{Li} = 0$ and $\varepsilon = 0$, Equations (4.11) and (4.12) become $\boldsymbol{w}(n+1) = \boldsymbol{w}(n)$.

Also, the alternative update relation for the sign regressor form of the maximum form of NLMS from (4.13) is written as

$$\boldsymbol{w}(n+1) = \boldsymbol{w}(n) + \frac{\mu'}{\varepsilon + \max\left(\|\boldsymbol{x}(n)\|\right)^2} \text{Sign}\{\boldsymbol{x}(n)\}\ e(n) \tag{4.14}$$

4.4 ERROR-NORMALIZED LMS (ENLMS) ALGORITHM

The squared error vector norm can be used rather than the instantaneous normalized data vector. The number of iterations is the error vector length. As step size is normalized with regard to error, this is known as ENLMS algorithm. In ENLMS technique, the time-varying step size is inversely proportional rather than the data vector as in the NLMS algorithm to the squared error vector norm. This technique also improves significantly in reducing distortion of signal [114]. Advantage of ENLMS is that the step size can also be chosen regardless of input signal power and amount of tap weights. ENLMS algorithm therefore provides a higher rate of convergence as well as steady-state error than LMS technique. The mathematical modeling of ENLMS algorithm is presented in Table 4.3.

The output of FIR filter is given by

$$y(n) = x^T(n)\ w(n) = w^T(n)\ x(n)$$

The expression for estimation error $e(n)$ is given by

$$e(n) = d(n) - w^T(n) \cdot x(n)$$

The relation for mean square error or cost function $J(n)$ is

$$J(n) = E\left[|d(n) - y(n)|^2\right] = E\left[|e(n)|^2\right]$$

The following expression results in minimizing $(e^*(n))^2$ in relation to $\beta(n)$:

$$\beta(n) = \frac{1}{2\ e^T(n)\ e(n)} \tag{4.17}$$

which forces $e^*(n)$ to be zero. The limit of the step size for mean-square LMS convergence is defined as

$$0 < \beta < \frac{2}{\boldsymbol{x}^T(n)\boldsymbol{x}(n)} \tag{4.19}$$

The constant used for adaptation β' is dimensionless for ENLMS filter, while the adaptation constant μ of LMS filter has inverse dimensions of power. The NLMS filter can be viewed as an LMS filter with a time-varying step parameter. The input vector $x(n)$ of the tap is small to overcome the LMS gradient noise amplification drawback for ENLMS algorithm.

There may be numerical problems, and then, it should be divided for the squared norm $\|e(n)\|^2$ by a small value. To prevent this issue, we alter the above-mentioned recursion by adding a small positive constant ε. The parameter ε is set to ensure that denominator is too small and step size is too large. The flow chart for the maximum form of ENLMS algorithm is as depicted in Figure 4.2.

Now the step size parameter is written as follows:

$$\beta(n) = \frac{\beta'}{\varepsilon + \|e(n)\|^2} \tag{4.21}$$

where $\beta(n)$ is a normalized step size with $0 < \beta' < 2$. Replacing μ in the LMS weight vector update equation with $\beta(n)$ from (4.20) leads to the ENLMS, which is given as

$$\boldsymbol{w}(n+1) = \boldsymbol{w}(n) + \frac{\beta'}{\|\boldsymbol{e}(n)\|^2} e(n)\boldsymbol{x}(n) \tag{4.22}$$

Comparing (3.17) and (4.23), the ENLMS weight update equation is a scaled variant of NLMS algorithm. Hence, variation in the size of weight vector $w(n)$ is therefore

TABLE 4.3
Mathematical Modeling of ENLMS Algorithm

Parameters: T = number of taps (i.e. filter length), μ = step size parameter
Let the tap input be $x(n)$ and filter length T is moderate to large.

Initialization: Set $w(0) = 0$ as the initial condition.

Data: Given $x(n)$ = T-by-1 tap input vector to filter n_2 at time n

$$= \left[x(n), x(n-1) \ldots x(n-T+1)\right]^T$$

$d(n)$ is desired response at time n, β' is the adaptation constant, error signal is defined as $e(n)$, ε is a small positive parameter to avoid denominator become too small, and $(.)^T$ is the transpose of (.).

To be computed:

$w(n+1)$ = estimate of tap-weight vector at time $n+1$

Computation:

$w(n)$ is the tap-weight vector with the taps stored in a row vector

$$= \left[w(n)\ w(n-1) \ldots w(n-T+1)\right]^T$$

The LMS recursion expression from (4.1) is given by

$$w(n+1) = w(n) + 2\beta(n)x(n)e(n) \tag{4.15}$$

where $\beta(n)$ is the time-varying step size selected to make posteriori error $e^*(n)$ in (4.2) is minimized in magnitude given

$$e^*(n) = \left(1 - 2\beta(n)x^T(n)x(n)e(n)\right) \tag{4.16}$$

From (4.17) in (4.15), the resultant expression is obtained as

$$w(n+1) = w(n) + \frac{1}{2e^T(n)e(n)}e(n)x(n) \tag{4.18}$$

Also, setting $\beta(n)$ in the case of ENLMS algorithm as

$$\beta(n) = \frac{\beta'}{e^T(n)e(n)} = \frac{\beta'}{\left\|e(n)\right\|^2} \tag{4.20}$$

A small positive constant parameter ε is added to prevent denominator from becoming too low, and hence, (4.20) becomes

$$\beta(n) = \frac{\beta'}{\varepsilon + \left\|e(n)\right\|^2} \tag{4.21}$$

where $\beta(n)$ is a normalized step size with $0 < \beta' < 2$.

By replacing β in the LMS weight vector update equation with $\beta(n)$ from (4.21), the weight update relation for ENLMS is written as

$$\boldsymbol{w}(n+1) = \boldsymbol{w}(n) + \frac{\beta'}{\varepsilon + \left\|e(n)\right\|^2}e(n)\boldsymbol{x}(n) \tag{4.23}$$

Thus, the weight update relation for maximum ENLMS for $e_{Li} \neq 0$ and $\varepsilon \neq 0$ becomes

$$\boldsymbol{w}(n+1) = \boldsymbol{w}(n) + \frac{\beta'}{\varepsilon + \max\left(\left\|e(n)\right\|\right)^2}e(n)\boldsymbol{x}(n) \tag{4.26}$$

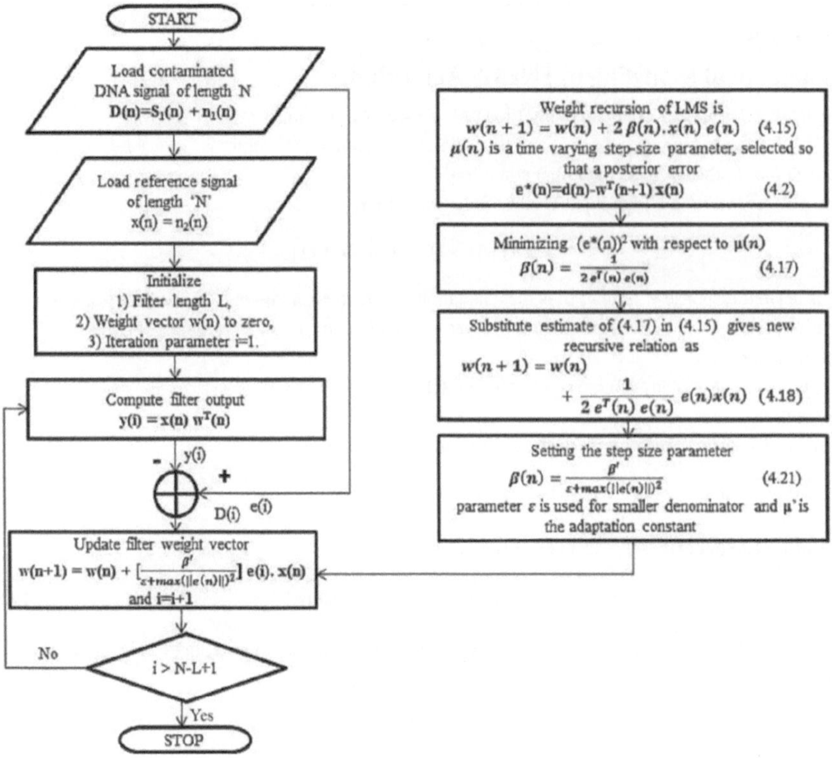

FIGURE 4.2 Flow chart of AEP based on maximum ENLMS algorithm.

inversely proportional with norm of the error vector $e(n)$. To achieve the best balance between convergence rate and low final mean square error $e(n)^2$, fixed parameter β is to be properly selected. The relation for $\beta(n)$ in (4.21) is a decreasing function of n with an increasing function of iteration parameter, n.

The ENLMS algorithm needs a small amount of calculations compared to other standardized algorithms [115]. To calculate the error value in (4.21) with minimal computational complexity, it is squared and stored in the first iteration. The second iteration error value is squared and added to the prior stored value. The outcome will then be saved to be used in the next iteration, and so on. On the other side, extra calculations are needed to calculate $\beta(n)$. In order to decrease computational difficulty, we have adopted block processing of ENLMS and its sign regressor version with overlapping blocks at this time in this section. The additional computations required to computing the variable step size for these error-normalized algorithms can be further reduced by using this block-based approach in which the input data is partitioned into blocks and the maximum magnitude within each block is used to compute variable step size parameter.

Now, the weight update relation of maximum ENLMS algorithm for $e_{Li} \neq 0$ and $\varepsilon = 0$ is written from (4.20) as

$$w(n+1) = w(n) + \frac{\beta'}{e_{Li}^2} x(n)\, e(n) \tag{4.24}$$

Here, we have also considered the weight update relation for the sign regressor form of maximum SRENLMS algorithm which is written from (4.20) as

$$w(n+1) = w(n) + \frac{\beta'}{e_{Li}^2} \text{Sign}\{x(n)\} e(n) \tag{4.25}$$

where $e_{Li} = \max\{|e_k|, k \in Z_i'\}$, $Z_i' = \{iL,\ iL+1, \dots, iL+L-1\}$, $i \in Z$, and for $e_{Li} = 0$ and $\varepsilon = 0$, Equations (4.11) and (4.12) become $w(n+1) = w(n)$.

Also, the alternative update relation for the sign regressor form of maximum ENLMS is written as

$$w(n+1) = w(n) + \frac{\beta'}{\varepsilon + \max(\|e(n)\|)^2} \text{Sign}\{x(n)\}\ e(n) \tag{4.27}$$

4.5 NORMALIZED LEAST MEAN FOURTH (NLMF) ALGORITHM

LMS technique suffers from such issues as amplification of gradient noise, low convergence, and weight drift. The normalized algorithm with the fourth-order minimization of estimation error is called NLMF algorithm [116]. Here, we have used NLMF and its adaptive algorithms based on SRA to enhance AEP efficiency. The NLMF algorithm overcomes the disadvantages of LMS and increases the ability of exon identification and quicker convergence when error is high. This also reduces the surplus EMSE in the exon identification process. These NLMF adaptive algorithms are used for developing AEPs in order to cope in real-time applications with the computational complexity of an AEP. The mathematical modeling of NLMF algorithm is presented in Table 4.4.

We have therefore created multiple AEPs, using NLMF algorithm and their variants, to minimize computational complexities and for quicker convergence. The benefits of the NLMF algorithm are improved stability and rapid convergence arising from normalization [117]. The step size of these algorithms is normalization pertaining to noise as well as signal power. If the length of tap is larger, the large tap duration in real-time applications creates an additional calculation burden of AEP.

Output of FIR filter is

$$y(n) = w^T(n)\ x(n) = x^T(n)\ w(n)$$

The expression for estimation error $e(n)$ is given by

$$e(n) = d(n) - y(n) = d(n) - w^T(n) \cdot x(n)$$

From the steepest descent algorithm, the expression for gradient vector is given by

$$\nabla J(n) = -2p + 2R\ w(n) = -2\ x(n) d^*(n) + 2\ x(n) x^T(n) w(n)$$

The expression for cost function of LMF is

$$J(n) = E\big[|e(n)|\big]^4$$

TABLE 4.4
Mathematical Modeling of NLMF Algorithm

Parameters: T = number of taps (i.e. filter length), μ = step size parameter
Let the tap input be $x(n)$ and filter length T is moderate to large.
Initialization: Set $w(0) = 0$ as the initial condition.
Data: Given $x(n) = T$-by-1 tap input vector to filter n_2 at time n

$$= [x(n), x(n-1) \ldots x(n-T+1)]^T$$

$d(n)$ is the desired response at time n, μ' is the adaptation constant, error signal is $e(n)$, ε is a small positive parameter to avoid denominator become too small, $e^*(n)$ is the posteriori error, and $(.)^T$ is the transpose of (.).
To be computed:
$w(n+1)$ = estimate of tap-weight vector at time $n+1$
Computation:
The Wiener solution for a linear estimation problem is given by

$$d(n) = w_0^T\ x(n) + e_0(n)$$

The LMS recursion expression from (4.1) is written as

$$w(n+1) = w(n) + 2\mu(n)x(n)e^3(n) \tag{4.28}$$

where $\mu(n)$ is the time variable step size parameter, chosen to minimize the extent of a posteriori error given by

$$e^*(n) = \left(1 - 2\ \mu(n)\ x^T(n)\ x(n)\ e^3(n)\right) \tag{4.29}$$

is minimized in magnitude.
From (4.4) in (4.28), the weight update expression is obtained as

$$w(n+1) = w(n) + \frac{1}{2x^T(n)x(n)} e^3(n)x(n) \tag{4.30}$$

Setting the time-varying size of step parameter as $\mu(n) = \dfrac{\mu'}{\mathbf{x}^T(n)\mathbf{x}(n)} = \dfrac{\mu'}{\|\mathbf{x}(n)\|^2}$

A small positive constant parameter ε is added to prevent denominator from becoming too low, and hence, (4.7) becomes

$$\mu(n) = \frac{\mu'}{\varepsilon + \|\mathbf{x}(n)\|^2} \tag{4.34}$$

By replacing μ in LMF weight equation in (3.24) with $\mu(n)$ expression from (4.8), weight relation for NLMF is written as

$$\mathbf{w}(n+1) = \mathbf{w}(n) + \frac{\mu'}{\varepsilon + \|\mathbf{x}(n)\|^2} e^3(n)\mathbf{x}(n) \tag{4.36}$$

Thus, the weight update relation for MNLMF for $x_{\mathrm{Li}} \neq 0$ and $\varepsilon \neq 0$ is written from (4.36) as

$$\mathbf{w}(n+1) = \mathbf{w}(n) + \frac{\mu'}{\varepsilon + \max\left(\|\mathbf{x}(n)\|\right)^2} e^3(n)x(n) \tag{4.39}$$

From (3.20), the resultant weight recursion expression of NLMF is written as

$$w(n+1) = w(n) + \mu(n)\, E\left\{e^3(n)x^*(n)\right\} \tag{4.31}$$

The expectation $E\left\{e^3(n)\, x^*(n)\right\}$ is generally unknown and must be replaced with an estimate such as sample mean given in (3.21).

Incorporating this estimation into the steepest decent, the weight update equation of NLMF becomes

$$w(n+1) = w(n) + \mu(n)\frac{1}{L}\sum_{l=0}^{L-1} e^3(n-l)x^*(n-l) \tag{4.32}$$

In LMS, for 1-point sample mean with $L = 1$ from (3.23), the weight recursion for the NLMF algorithm is derived as

$$w(n+1) = w(n) + \mu(n)\, e^3(n)\, x(n) \tag{4.33}$$

$\mu(n)$ is a normalized step size with $0 < \mu' < 2$.

The adaptation constant μ' for the NLMF filter is dimensionless, and a small parameter ε is set to avoid denominator being too small and step size parameter being too big [118].

Replacing μ in the LMF weight vector update equation in (3.24) with $\mu(n)$ expression from (4.7) leads to the NLMF, which is given as

$$\boldsymbol{w}(n+1) = \boldsymbol{w}(n) + \frac{\mu'}{\left\|\boldsymbol{x}(n)\right\|^2} e^3(n)\, \boldsymbol{x}(n) \tag{4.35}$$

With the normalization of the LMS step size by $\left\|\boldsymbol{x}(n)\right\|^2$ in the NLMF algorithm, however, this noise amplification problem is diminished. The flow chart for the maximum form of NLMF algorithm is as depicted in Figure 4.3.

To reduce the computational complexity, we have adopted block processing of NLMF and its sign regressor version with overlapping blocks. The additional computations required to computing the variable step size for these error-normalized algorithms can be further reduced by using this block-based method in which input data is partitioned into blocks and the maximum magnitude within each block is used to compute variable step size parameter. This method is also termed as maximum normalization.

Now, the weight update relation of maximum NLMF (MNLMF) algorithm for $x_{Li} \neq 0$ and $\varepsilon = 0$ is written from (4.35) as

$$\boldsymbol{w}(n+1) = \boldsymbol{w}(n) + \frac{\mu'}{x_{Li}^2}\, e^3(n)\boldsymbol{x}(n) \tag{4.37}$$

Here, we have also considered the weight update relation for the sign regressor form of maximum SRNLMF algorithm which is written from (4.35) as

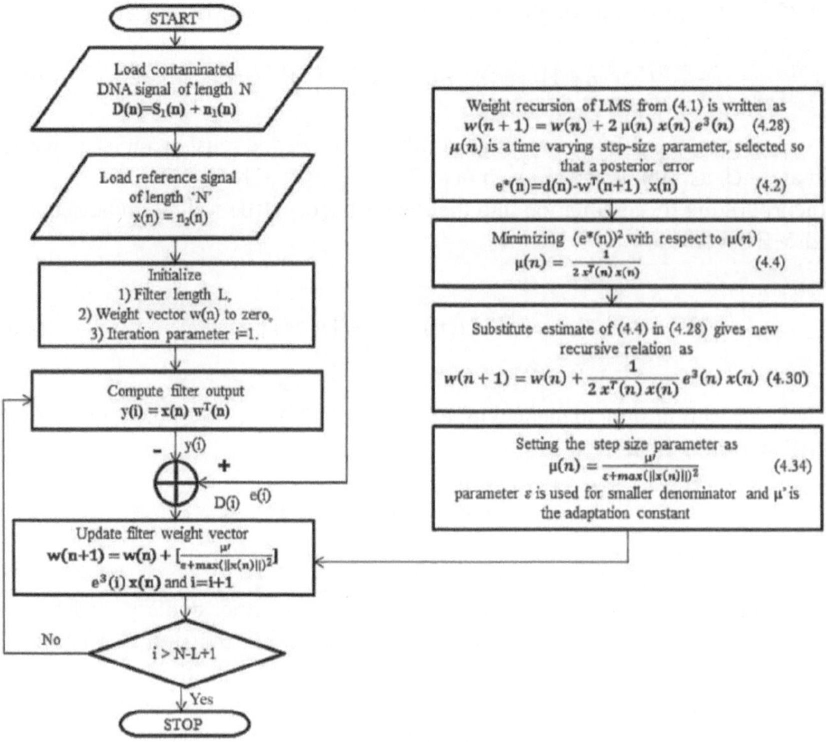

FIGURE 4.3 Flow chart of AEP based on MNLMF algorithm.

$$w(n+1) = w(n) + \frac{\mu'}{x_{Li}^2} e^3(n)\ \text{Sign}\{x(n)\} \tag{4.38}$$

where $x_{Li} = \max\{|x_k|, k \in Z_i'\}$, $Z_i' = \{iL,\ iL+1, \ldots, iL+L-1\}$, $i \in Z$, and for $x_{Li} = 0$ and $\varepsilon = 0$, Equations (4.11) and (4.12) become $w(n+1) = w(n)$.

Also, the alternative update relation for sign regressor form of MNLMF is written as

$$w(n+1) = w(n) + \frac{\mu'}{\varepsilon + \max(x(n))^2} e^3(n)\ \text{Sign}\{x(n)\} \tag{4.40}$$

4.6 VARIABLE STEP SIZE NORMALIZED LMS (VNLMS) ALGORITHM

This technique with a varying step size is regarded to enhance AEP performance over LMS-based algorithms and other normalized variants. The fixed step size algorithms are independent of data and step size variations do not follow the tracking criteria. To overcome this setback, an NLMS algorithm with a variable step size and its signed variants are used [119]. VNLMS algorithm overcomes the disadvantages of LMS and increases the speed of convergence and tracking capability. For variable step size algorithms, there are many recommendations.

We have considered the implementation of our AEP with use of variable step strategy because of its faster convergence and good exon tracking ability compared to LMS algorithm. Combining this variable step size with NLMS results in VNLMS algorithm [120]. VNLMS is more beneficial than LMS due to its stability achieved by normalization. To reduce the computational complexity, we have combined VNLMS algorithm with sign regressor algorithm. The resulting algorithm is variable normalized SRLMS (VNSRLMS) algorithm which offers good tracking ability and low computational complexity. The mathematical modeling of NLMS algorithm is presented in Table 4.5.

The output of FIR filter is given by

$$y(n) = x^T(n)\, w(n) = w^T(n)\, x(n)$$

The expression for estimation error $e(n)$ is given by

$$e(n) = d(n) - w^T(n) \cdot x(n)$$

The relation for mean square error or cost function $J(n)$ is

$$J(n) = E\left[|d(n) - y(n)|^2\right] = E\left[|e(n)|^2\right]$$

From (3.26), the resultant weight recursion expression of NVSLMS is written as

$$w(n+1) = w(n) + \mu'(n)\, E\{e(n)x^*(n)\} \tag{4.44}$$

The expectation $E\{e^3(n)\, x^*(n)\}$ is generally unknown and must be replaced with an estimate such as the sample mean given in (3.21).

Incorporating this estimation into the steepest decent, the weight update equation of VNLMS becomes

$$w(n+1) = w(n) + \mu'(n)\frac{1}{L}\sum_{l=0}^{L-1} e(n-l)x^*(n-l) \tag{4.45}$$

In LMS, for 1-point sample mean with $L = 1$, the weight recursion for the VNLMS algorithm is derived as

$$w(n+1) = w(n) + \mu'(n)\, e(n)\, x(n) \tag{4.46}$$

The normalized varying step size parameter $\mu'(n)$ is set as given in (4.47). The adaptation constant $\mu(n)$ for the VNLMS filter is dimensionless. The VNLMS filter presents its own issue in overcoming the gradient noise amplification issue associated with standard LMS filter, namely the tap input vector $x(n)$ is small [121].

Numerical problems may occur here, and then we must divide by a small value for the squared norm $\|x(n)\|^2$. To resolve this problem, we alter the previous recursion by adding a small positive constant ε. It is set to prevent denominator being too small and step size parameter being too large ε.

TABLE 4.5
Mathematical Modeling of VNLMS Algorithm

Parameters: T = number of taps (i.e. filter length), μ = step size parameter
Let the tap input be $x(n)$ and filter length T is moderate to large.
Initialization: Set $w(0) = 0$ as the initial condition.
Data: Given $x(n) = T$-by-1 tap input vector to filter n_2 at time n

$$= \left[x(n), x(n-1) \ldots x(n-T+1)\right]^T$$

$d(n)$ is the desired response at time n, μ' is the adaptation constant, error signal is $e(n)$, ε is a small positive parameter to avoid denominator become too small, $e^*(n)$ is the posteriori error, and $(.)^T$ is the transpose of (.).
To be computed: $w(n+1)$ = estimate of tap-weight vector at time $n+1$
Computation:
$w(n)$ is tap-weight vector with taps stored in a row vector

$$= \left[w(n)\ w(n-1) \ldots\ w(n-T+1)\right]^T$$

The LMS recursion expression from (4.1) is written as

$$w(n+1) = w(n) + 2\mu'(n)x(n)e(n) \tag{4.41}$$

where $\mu'(n)$ is the normalized step size parameter in (4.41) selected so that the posteriori error given by

$$e^*(n) = \left(1 - 2\mu'(n)x^T(n)x(n)e(n)\right) \tag{4.42}$$

is minimized in magnitude.
From (4.4) in (4.41), the weight update expression is obtained as

$$w(n+1) = w(n) + \frac{1}{2x^T(n)x(n)}e(n)x(n) \tag{4.43}$$

Setting $\mu'(n)$ in the case of VNLMS algorithm, it becomes

$$\mu(n) = \frac{\mu'}{\boldsymbol{x}^T(n)\boldsymbol{x}(n)} = \frac{\mu'}{\|\boldsymbol{x}(n)\|^2} \tag{4.47}$$

A small positive constant parameter ε is added to prevent denominator from becoming too low, and hence, (4.7) becomes

$$\mu(n) = \frac{\mu'}{\varepsilon + \|\boldsymbol{x}(n)\|^2} \tag{4.48}$$

where $\mu'(n)$ is a normalized step size parameter with $0 < \mu(n) < 2$.
Thus, the variable step size parameter $\mu(n)$ of VSLMS algorithm is calculated as

$$\mu(n) = \mu_{\max} + (\mu_{\min} - \mu_{\max})e^{-\beta}\sigma^2 e^{(n)} \tag{4.49}$$

By replacing μ in VSLMS weight equation in (3.28) with $\mu(n)$ expression from (4.48), weight relation for VNLMS is written as

$$\boldsymbol{w}(n+1) = \boldsymbol{w}(n) + \frac{\mu(n)}{\varepsilon + \|\boldsymbol{x}(n)\|^2}e(n)\boldsymbol{x}(n) \tag{4.51}$$

Thus, the weight update relation for maximum VNSRLMS (MVNSRLMS) for $x_{Li} \neq 0$ and $\varepsilon \neq 0$ is written from (4.51) as

$$\boldsymbol{w}(n+1) = \boldsymbol{w}(n) + \frac{\mu(n)}{\varepsilon + \max\left(\|\boldsymbol{x}(n)\|\right)^2}e(n)x(n) \tag{4.54}$$

In (4.49), $0 < \mu_{\min} < \mu_{\max}$ is chosen to provide minimum capability of tracking as well as algorithm stability. The parameter β is a positive number that adds some flexibility of algorithm design and $\sigma^2 e^{(n)}$ is a power of instantaneous error. These fluctuations are due to noisy gradient vectors used to make corrections to $w(n)$.

Replacing μ in the VSLMS weight vector update equation in (3.28) with $\mu'(n)$ expression from (4.47) contributes to NVSLMS, indicated as

$$\boldsymbol{w}(n+1) = \boldsymbol{w}(n) + \frac{\mu(n)}{\|\boldsymbol{x}(n)\|^2} e(n)\, \boldsymbol{x}(n) \tag{4.50}$$

The correction in the LMS algorithm $w(n)$ is large proportional to the input vector $x(n)$. The LMS algorithm is problematic with gradient noise amplification. This noise amplification issue is, however, reduced with the normalization of the LMS step size by $\|\boldsymbol{x}(n)\|^2$ in VNLMS algorithm [122].

Although the VNLMS algorithm bypasses the noise amplification problem, we face a similar problem when $\|\boldsymbol{x}(n)\|$ is too low.

The flow chart for the maximum form of VNLMS algorithm is as depicted in Figure 4.4. Due to normalization, the length of long filter tap can be reduced to one without regard to tap length by means of an approach known as maximum variable normalization. To reduce the computational complexity, we have adopted

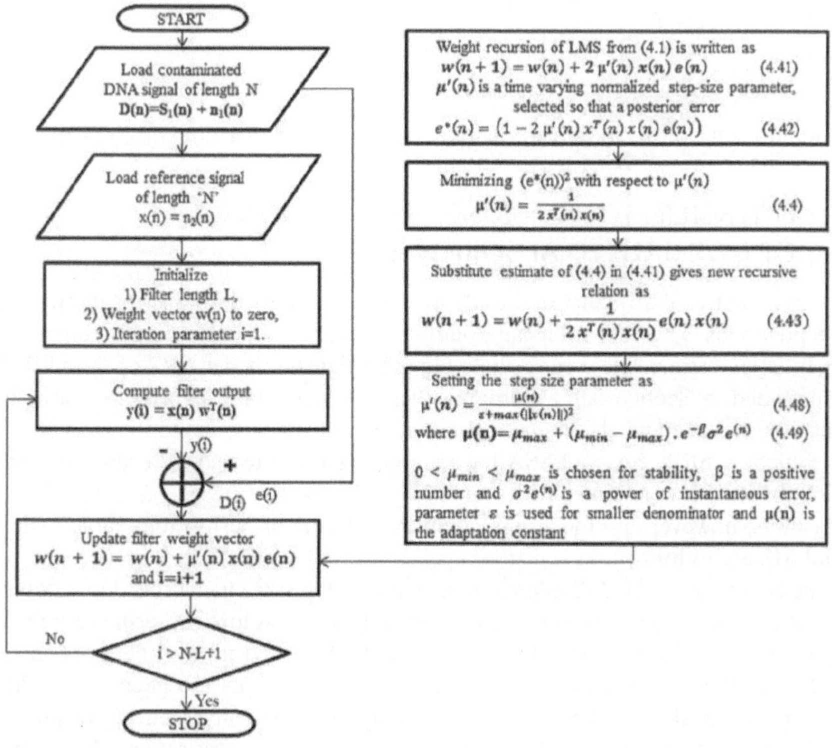

FIGURE 4.4 Flow chart of AEP based on maximum VNLMS algorithm.

block processing of VNLMS and its sign-based versions with overlapping blocks. The additional computations required to computing the variable step size for these variable normalized algorithms can be further reduced by using this block-based method with the input data divided into blocks and the maximum magnitude within each block is used to compute variable step size parameter. Now the weight update relation of maximum VNLMS (MVNLMS) algorithm for $x_{Li} \neq 0$ and $\varepsilon = 0$ is written from (4.50) as

$$w(n+1) = w(n) + \frac{\mu(n)}{x_{Li}^2} e(n)\, x(n) \tag{4.52}$$

Here, we have also considered the weight update relation for the sign regressor form of maximum VNSRLMS (MVNSRLMS) algorithm which is written from (4.50) as

$$w(n+1) = w(n) + \frac{\mu(n)}{x_{Li}^2} e(n)\, \mathrm{Sign}\{w(n)\} \tag{4.53}$$

where $x_{Li} = \max\{|x_k|, k \in Z_i'\}, Z_i' = \{iL,\ iL+1, \ldots, iL+L-1\}, i \in Z$, and for $x_{Li} = 0$ and $\varepsilon = 0$, Equations (4.52) and (4.53) become $w(n+1) = w(n)$.

Also, the alternative update relation for the sign regressor form of MVNSRLMS is written from (4.54) as

$$w(n+1) = w(n) + \frac{\mu(n)}{\varepsilon + \max(\|x(n)\|)^2} e(n)\, \mathrm{Sign}\{x(n)\} \tag{4.55}$$

where $\mu(n)$ is the variable step size parameter of VSLMS algorithm as in (4.49).

4.7 EXTENSION TO SIGN-BASED REALIZATIONS OF NORMALIZED ALGORITHMS

Less computational difficulty of the adaptive filter is highly desirable in bioinformatics applications. This reduction can usually be achieved either by clipping input data or estimation errors. The error or data clipping algorithms, i.e. SRA, SA and SSA, are provided in Section 3.9. Combining these with standard algorithms, they give quick convergence and also reduces computational difficulty due to the existence of sign function. SRA, SA, and SSA have a convergence rate and a steady-state error which for the same parameter setting is slightly less than LMS algorithms. These algorithms, however, are much less complex in terms of performing computations than LMS algorithm.

The advantage of NLMS algorithm is that the step size can be selected independent of the input signal power and the tap weight. Therefore, in terms of quicker convergence rate and steady-state error, the NLMS algorithm is better than the LMS algorithm. On the other side, additional calculations are needed to calculate $\mu(n)$. To deal with the problems of complexity or convergence, we present NLMS and its sign regressor–based normalized algorithms for accurately locating exon segments in gene sequences such as the signed regressor–normalized LMS (SR-NLMS) algorithm [123].

4.7.1 Sign-Based Normalized LMS (NLMS) Algorithms

NLMS algorithm is a basic version of higher-order adaptive filter. Combining the NLMS in (4.10) with SRA, SA, and SSA results in its signed variants NSRLMS, NSLMS, and NSSLMS, respectively. The updated weight expressions for these algorithms are provided as follows:

$$\boldsymbol{w}(n+1)=\boldsymbol{w}(n)+\frac{\mu'}{\varepsilon+\|\boldsymbol{x}(n)\|^2}e(n)\ \boldsymbol{Sign}\{\boldsymbol{x}(n)\} \tag{4.56}$$

$$\boldsymbol{w}(n+1)=\boldsymbol{w}(n)+\frac{\mu'}{\varepsilon+\|\boldsymbol{x}(n)\|^2}\mathrm{Sign}\{e(n)\}\boldsymbol{x}(n) \tag{4.57}$$

$$\boldsymbol{w}(n+1)=\boldsymbol{w}(n)+\frac{\mu'}{\varepsilon+\|\boldsymbol{x}(n)\|^2}\mathrm{Sign}\{e(n)\}\boldsymbol{Sign}\{\boldsymbol{x}(n)\} \tag{4.58}$$

From (4.57) and (4.58), it is noted that the weight update relations for NSLMS and NSSLMS are the same as SLMS and SSLMS. Among these, the NSRLMS algorithm in (4.56) performs better because of its low computational complexity, better convergence performance, and good tracking ability. To further decrease the computation complexity, we use overlapping blocks to adopt sign-based NLMS algorithms. The resulting algorithms include MNLMS, MNSRLMS, MNSLMS, and MNSSLMS.

The weights update relations of MNLMS and MNSRLMS algorithms for $x_{Li} \neq 0$ and $\varepsilon \neq 0$ are given as in (4.13) and (4.14), respectively.

Similarly, the weight update equations for MNSLMS and MNSSLMS are written as

$$\boldsymbol{w}(n+1)=\boldsymbol{w}(n)+\frac{\mu'}{\varepsilon+\max\left(\|\boldsymbol{x}(n)\|\right)^2}x(n)\ \mathrm{Sign}\{\boldsymbol{e}(n)\} \tag{4.59}$$

$$\boldsymbol{w}(n+1)=\boldsymbol{w}(n)+\frac{\mu'}{\varepsilon+\max\left(\|\boldsymbol{x}(n)\|\right)^2}\mathrm{Sign}\{\boldsymbol{x}(n)\}\ \mathrm{Sign}\{e(n)\} \tag{4.60}$$

Among these normalized and its maximum variants, MNSRLMS algorithm in (4.14) performs better than LMS and all its other variants due to its low computational complexity and good exon tracking ability.

4.7.2 Sign-Based Error-Normalized LMS (ENLMS) Algorithms

ENLMS algorithm is a basic version of higher-order adaptive filter. Combining the ENLMS in (4.23) with SRA, SA, and SSA results in its signed variants ENSRLMS,

ENSLMS, and ENSSLMS, respectively. For such algorithms, the weight update expressions are provided as follows:

$$\boldsymbol{w}(n+1)=\boldsymbol{w}(n)+\frac{\mu'}{\varepsilon+\|\boldsymbol{e}(n)\|^2}e(n)\ \boldsymbol{Sign}\{\boldsymbol{x}(n)\} \tag{4.61}$$

$$\boldsymbol{w}(n+1)=\boldsymbol{w}(n)+\frac{\mu'}{\varepsilon+\|\boldsymbol{e}(n)\|^2}\mathrm{Sign}\{e(n)\}\ \boldsymbol{x}(n) \tag{4.62}$$

$$\boldsymbol{w}(n+1)=\boldsymbol{w}(n)+\frac{\mu'}{\varepsilon+\|\boldsymbol{e}(n)\|^2}\mathrm{Sign}\{e(n)\}\boldsymbol{Sign}\{\boldsymbol{x}(n)\} \tag{4.63}$$

From (4.62) and (4.63), it is noted that the weight update relations for ENSLMS and ENSSLMS are the same as SLMS and SSLMS. To further decrease the computational complexity, we use overlapping blocks to process sign-based ENLMS algorithms. The resulting algorithms include MENLMS, MENSRLMS, MENSLMS, and MENSSLMS.

The weights update relations of MENLMS and MENSRLMS algorithms for $x_{Li} \neq 0$ and $\varepsilon \neq 0$ are given as in (4.26) and (4.27), respectively.

Similarly, weight update equations for MENSLMS and MENSSLMS are written as

$$\boldsymbol{w}(n+1)=\boldsymbol{w}(n)+\frac{\beta'}{\varepsilon+\max\left(\|\boldsymbol{e}(n)\|\right)^2}\boldsymbol{x}(n)\ \mathrm{Sign}\{e(n)\} \tag{4.64}$$

$$\boldsymbol{w}(n+1)=\boldsymbol{w}(n)+\frac{\beta'}{\varepsilon+\max\left(\|\boldsymbol{e}(n)\|\right)^2}\mathrm{Sign}\{\boldsymbol{x}(n)\}\mathrm{Sign}\{e(n)\} \tag{4.65}$$

Among these error-normalized and its maximum variants, the MENSRLMS algorithm in (4.27) performs better than LMS and all its other variants due to its low computational complexity and good exon tracking ability.

4.7.3 Sign-Based Normalized LMF (NLMF) Algorithms

NLMF algorithm is a basic version of higher-order adaptive filter. Combing the NLMF in (4.36) with SRA, SA, and SSA results in its signed variants NSRLMF, NSLMF, and NSSLMF, respectively. For such algorithms, the weight update expressions are as follows:

$$\boldsymbol{w}(n+1)=\boldsymbol{w}(n)+\frac{\mu'}{\varepsilon+\|\boldsymbol{x}(n)\|^2}e^3(n)\boldsymbol{Sign}\{\boldsymbol{x}(n)\} \tag{4.66}$$

$$\boldsymbol{w}(n+1)=\boldsymbol{w}(n)+\frac{\mu'}{\varepsilon+\|\boldsymbol{x}(n)\|^2}\mathrm{Sign}\{e^3(n)\}\ \boldsymbol{x}(n) \tag{4.67}$$

$$\boldsymbol{w}(n+1)=\boldsymbol{w}(n)+\frac{\mu'}{\varepsilon+\|\boldsymbol{x}(n)\|^2}\mathrm{Sign}\{e^3(n)\}\ \boldsymbol{Sign}\{\boldsymbol{x}(n)\} \tag{4.68}$$

From (4.67) and (4.68), it is noted that the weight update relations for NSSLMF and NSSLMF are the same as SLMS and SSLMS. In order to reduce computing complexity further, we use overlapping blocks to process sign-based ENLMS algorithms. The resulting algorithms include MNLMF, MNSRLMF, MNSLMF, and MNSSLMF.

The weight update relations of MNLMF and MNSRLMF algorithms for $x_{Li} \neq 0$ and $\varepsilon \neq 0$ are written as in (4.39) and (4.40), respectively.

Similarly, for MNSLMF and MNSSLMF, weight relations are

$$w(n+1) = w(n) + \frac{\mu'}{\varepsilon + \max\left(\|x(n)\|\right)^2} \text{Sign}\left\{e^3(n)\right\} x(n) \tag{4.69}$$

$$w(n+1) = w(n) + \frac{\mu'}{\varepsilon + \max\left(\|x(n)\|\right)^2} \text{Sign}\left\{e^3(n)\right\} \text{Sign}\left\{x(n)\right\} \tag{4.70}$$

Among these normalized and maximum variants, MNSRLMF algorithm in (4.40) performs better than LMS and all its other variants due to its low computational complexity and good exon tracking ability.

4.7.4 Sign-Based Variable Step Size NLMS (VNLMS) Algorithms

VNLMS algorithm is a basic version of higher-order adaptive filter. Combining the VNLMS in (4.51) with SRA, SA, and SSA results in its signed variants VNSRLMS, VNSLMS, and VNSSLMS algorithms, respectively. For such algorithms, the weight update expressions are as follows:

$$w(n+1) = w(n) + \frac{\mu(n)}{\varepsilon + \|x(n)\|^2} e(n) \text{ Sign}\left\{x(n)\right\} \tag{4.71}$$

$$w(n+1) = w(n) + \frac{\mu(n)}{\varepsilon + \|x(n)\|^2} \text{Sign}\left\{e(n)\right\} x(n) \tag{4.72}$$

$$w(n+1) = w(n) + \frac{\mu(n)}{\varepsilon + \|x(n)\|^2} \text{Sign}\left\{e(n)\right\} x(n) \tag{4.73}$$

From (4.72) and (4.73), it is noted that the weight update relations for VNSLMS and VNSSLMS are the same as SLMS and SSLMS. To further decrease the computational complexity, we embrace sign-based block processing of NLMS algorithms by considering overlapping blocks. The resulting algorithms include MVNLMS, MVNSRLMS, MVNSLMS, and MVNSSLMS.

The weight update relations of MVNLMS and MVNSRLMS algorithms for $x_{Li} \neq 0$ and $\varepsilon \neq 0$ are written as in (4.54) and (4.55), respectively.

Similarly, weight update equations for MVNSLMS and MVNSSLMS are written as

$$w(n+1) = w(n) + \frac{\mu(n)}{\varepsilon + \max\left(\|x(n)\|\right)^2} \text{Sign}\left\{x(n)\right\} e(n) \tag{4.74}$$

$$w(n+1) = w(n) + \frac{\mu(n)}{\varepsilon + \max\left(\|x(n)\|\right)^2} \operatorname{Sign}\{x(n)\} \operatorname{Sign}\{e(n)\} \qquad (4.75)$$

Among these normalized and maximum variants, the MVNSRLMS algorithm in (4.55) performs better than LMS and all its other variants due to its low computational complexity and good exon tracking ability.

4.8 COMPUTATIONAL COMPLEXITY ISSUES

In this section, we deliberate on comparing the computational complexities among distinct adaptive techniques based on normalization. Computational complexity values necessary to calculate all normalized adaptive techniques and its variants based on sign function are presented in Table 4.6. The signed accomplishments significantly lessen the amount of calculations needed for NLMS algorithm. Furthermore, these algorithms based on SRA are mainly free of multiplication operations; these algorithms provide elegant means for genomic sequence analysis. Among these normalized algorithms, complexity involved in NLMS technique is more because it requires 2T+1 multiplication and addition operations to calculate the equation for the weight recursion in (4.10), where T is the filter length. Similarly, ENLMS also requires 2T+1 multiplication and addition operations. Only single division is necessary to calculate

TABLE 4.6
Computational Complexity Comparison of Normalized and Maximum Normalized Adaptive Algorithms

S.		Without Maximum Normalization			With Maximum Normalization		
No.	Algorithm	Multiplications	Additions	Divisions	Multiplications	Additions	Divisions
1	NLMS	2T+1	2T+1	1	T+1	T	1
2	NSRLMS	2T+1	T+1	1	2	T+1	1
3	NSLMS	2T+1	2T	1	T+1	T+1	1
4	NSSLMS	2T+1	T+1	1	1	2	1
5	ENLMS	2T+1	2T+1	1	T+1	T	1
6	ENSRLMS	2T+1	T+1	1	2	T+1	1
7	ENSLMS	2T+1	2T	1	T+1	T+1	1
8	ENSSLMS	2T+1	T+1	1	1	2	1
9	NLMF	2T+3	2T+1	1	T+3	T+1	1
10	NSRLMF	2T+1	T+3	1	5	T+1	1
11	NSLMF	2T+1	2T	1	T+1	T+1	1
12	NSSLMF	2T+1	T+1	1	1	2	1
13	VNLMS	2T+1	2T+1	1	T+1	T	1
14	VNSRLMS	2T+1	T+1	1	2	T+1	1
15	VNSLMS	2T+1	2T	1	T+1	T+1	1
16	VNSSLMS	2T+1	T+1	1	1	2	1

$\mu \cdot e(n)$ in SRA, while there are 2T+1 multiplications required for two other signed NLMS algorithms if chosen a μ value for a power of two.

In such instances, multiplication becomes a shift operation that is less complicated in practical situations. In SSA, signum function is used for both the data and vector; also add μ to the weight vector using addition with sign check (ASC) operation. It is evident from Table 4.6 that the number of calculations needed for NSRLMS and ENSRLMS algorithms is independent of filter length. Both these algorithms required a less number of multiplications and one division, because of the sign term applied to data vector, i.e. all the elements of the data vector become −1 or 0 or 1, so that multiplications are greatly reduced.

The LMF algorithm considerably improves to reduce signal distortion. It is evident that NLMF algorithm requires 2T+3 multiplications and zero addition operations, whereas VNLMS algorithm requires only 2T+1 multiplication operations with a similar amount of additions. However, both the NLMF and VNLMS algorithms offer low computational complexity compared to all other LMS-based variants. Among these two algorithms, VNLMS offers low complexity in performing the computations and is more suitable in real bioinformatics-based applications.

The NSRLMF algorithm clearly enables the exons to be accurately located in PSD plot. The parameter α is used to regulate the convergence effectively. When there is a large error, α tends towards unity and convergence is very quick. Similarly, when error is low, α is small and convergence is slow to reduce the step size. This actually happens when the adaptive filter is stable. To further improve the convergence performance, VNLMS algorithm is regarded along with its signed variants.

To decrease the computational difficulty further, we have also considered signed maximum variants of all normalization-based adaptive algorithms as shown in Table 4.6. In the case of maximum variants, the number of computations required is greatly reduced compared to normalized variants; therefore, less multiplication and addition operations are needed to compute weight equation. From these computations, it is vibrant that VNSRLMS-based algorithm requires a lesser number of multiplications compared to all other signed LMS-based algorithms. Hence, AEP based on maximum VNSRLMS may be better suited for clinical remote healthcare monitoring systems and bioinformatics systems.

As a consequence of the clipping, only one multiplication is essential to a simple LMS combined with SRA, while maximum VNSRLMS-based AEP requires fewer multiplications and is independent of filter length. The lesser calculations for maximum VNSRLMS-based AEP are due to the sign function involved in the weight update expression. But the quality of the signal will be degraded by clipping the data and error vectors. Thus, in signal processing applications, VNSSLMS could not be a better candidate. Based on these factors and a lesser number of calculations, maximum VNSRLMS is definitely a better candidate for reliable prediction of exons in DNA sequence in genomic applications.

Among all the normalized based and their signed adaptive algorithms, all the algorithms perform better than LMS algorithms, whereas maximum VNSRLMS algorithm outperforms all the normalized variants in terms of low computational complexity and better tracking ability. To significantly reduce the computational complexity, we have also extended for block processing of all normalized adaptive algorithms and their sign regressor versions with overlapping blocks. Their resulting maximum normalized based adaptive algorithms include MNLMS, MNSRLMS, MNSLMS, MNSSLMS, MENLMS, MENSRLMS, MENSLMS, MENSSLMS, MMNLMF, MNSRLMF, MNSLMF, MNSSLMF, MVNLMS, MVNSRLMS, MVNSLMS, and MVNSSLMS. Among all these maximum normalized based and their signed adaptive algorithms, all the algorithms perform better than LMS algorithms, whereas MVNSRLMS algorithm outperforms all the normalized variants in terms of low computational complexity and better tracking ability.

4.9 CONVERGENCE ANALYSIS

Convergence curves between MSE and the number of iterations are plotted. Computation of MSE is done using the expression, i.e. $J = E\left\{\left|e(n)^2\right|\right\}$, in (3.8). The simulations for all the convergence curves are done using MATLAB. In this code, MSE is calculated for 4000 iterations for each sample, and the average value for the characterization is also taken. These curves are obtained using various normalized based AEPs for exon prediction using an adaptive finite length filter with a random variance of 0.01 and a step size of 0.01. The convergence curves for various adaptive normalized algorithms mentioned in the earlier sections are shown in Figure 4.5. From the convergence plots, it is evident that VNLMS outperforms among all the normalized based adaptive algorithms. Thus, the performances of all NLMS, ENLMS, NLMF, and VNLMS algorithms are comparable and superior to each other and LMS.

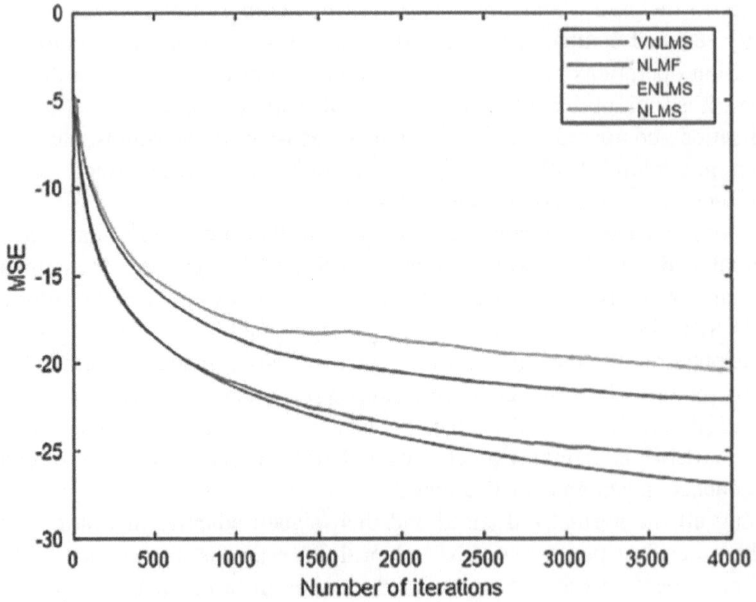

FIGURE 4.5 Convergence curves of normalized based adaptive algorithms.

Due to the normalization factor, the performance of sign regressor version of all NLMS algorithms is slightly inferior to the corresponding non-signed version of the algorithm; the corresponding convergence curves are shown in Figure 4.5. Due to its lower computational complexity, sign regressor NLMS–based algorithm offers better exon tracking ability and is suitable for applications in real time. Comparing the convergence performances of all normalized algorithms with LMS algorithm in Figure 3.10, it is evident that all normalized algorithms provide quicker convergence. The NLMF algorithm substantially improves the way signal distortion is minimized. It is obvious that the VNLMS algorithm converges faster than LMS and NLMS algorithms. The parameter α is used to actually control the convergence. If a large error is occurred, then α tends to unity and results in faster convergence.

Likewise, α is small when the error is low as well as convergence is slow to reduce the step size. This occurs when the adaptive filter reaches its steady state. To further improve the convergence performance, VNSLMS adaptive algorithm is considered. Among all normalized based algorithms, it is evident from Figure 4.5 that VNLMS offers better convergence performance compared to LMS and other normalized based algorithms. The convergence curves result from plotting the MSE over several samples.

To significantly reduce the computational complexity and for better convergence performance, we have further extended all these normalized algorithms to maximum normalized based adaptive algorithms. The resulting algorithms include MNLMS, MENLMS, MNLMF, and MVNLMS algorithms and their sign-based variants provided in the previous section. Figures 4.6–4.10 show the convergence

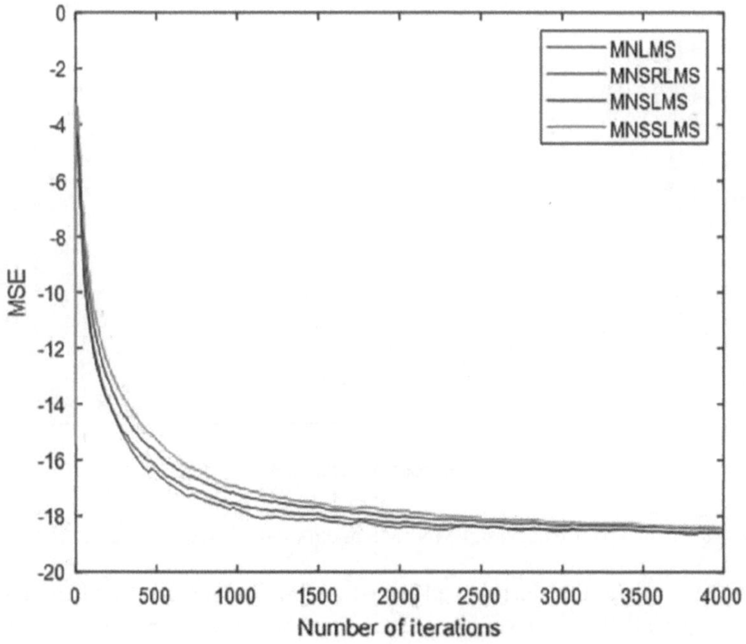

FIGURE 4.6 Convergence curves of MNLMS and its signed adaptive algorithms.

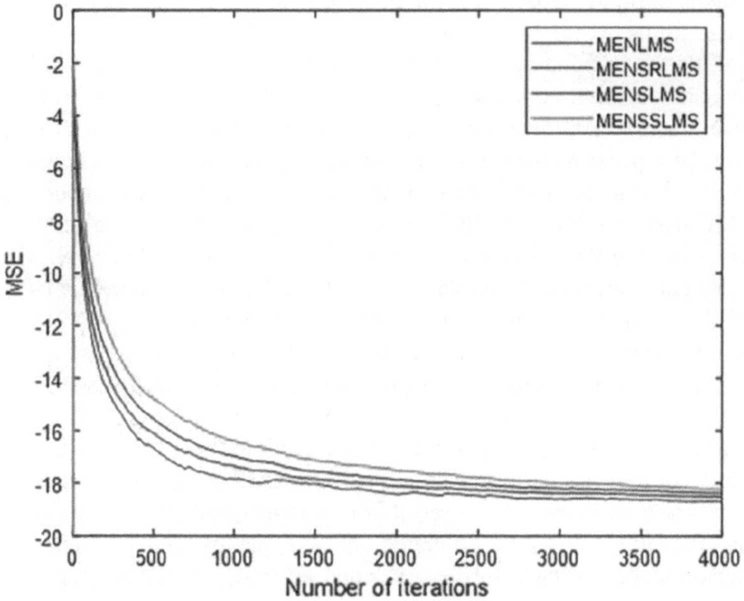

FIGURE 4.7 Convergence curves of MENLMS and its signed adaptive algorithms.

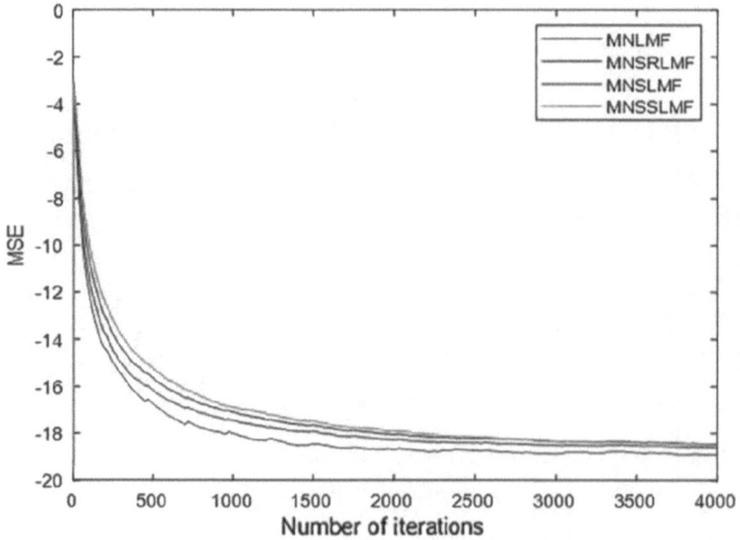

FIGURE 4.8 Convergence curves of MNLMF and its signed adaptive algorithms.

curves for all maximum normalized based adaptive algorithms and their signed variants. From all these curves, it is clear that the convergence performances of MNLMS, MENLMS, MNLMF, and MVNLMS algorithms converge faster than LMS by adopting normalization as it utilizes a variable convergence factor to minimize instantaneous output error.

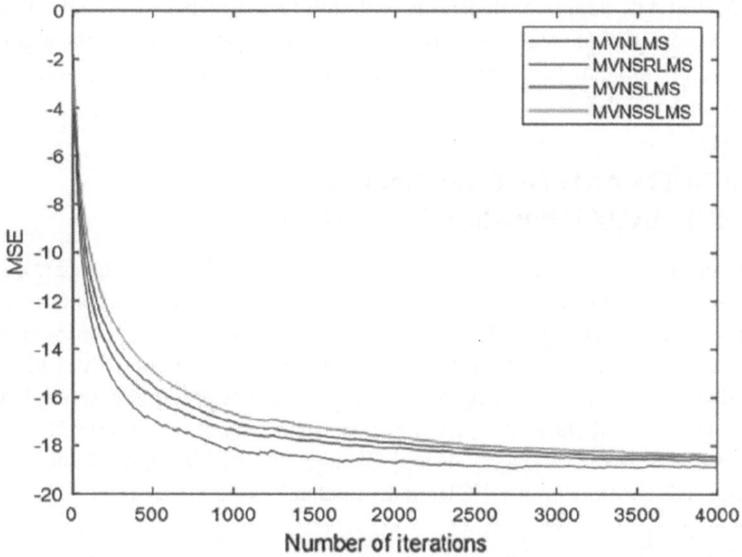

FIGURE 4.9 Convergence curves of MVNLMS and its signed adaptive algorithms.

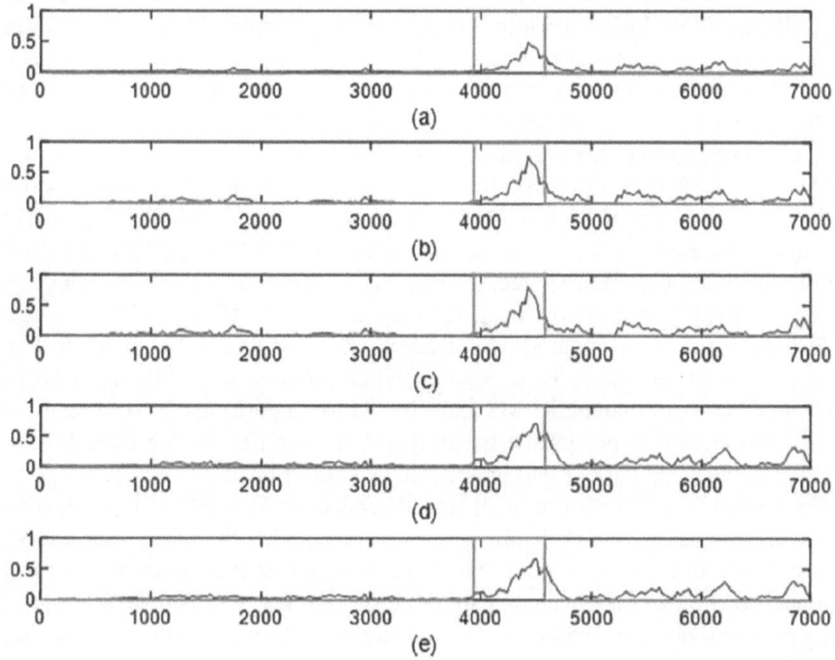

FIGURE 4.10 PSD plots with the location of exon (3934-4581) for a DNA sequence with accession AF009962 predicted using (a) LMS-based AEP, (b) MNLMS-based AEP, (c) MNSRLMS-based AEP, (d) MNSLMS-based AEP, and (e) MNSSLMS-based AEP. (Relative base location is taken on x-axis, and power spectrum is taken on y-axis.)

The MVNLMS adaptive algorithm is therefore better based on computational complexity, convergence features, and exon locating capability among the algorithms considered for implementing AEPs than all other maximally normalized adaptive algorithms.

4.10 RESULTS AND DISCUSSION FOR NORMALIZATION-BASED VARIANTS

Various performances of maximum normalized based AEPs are compared in this chapter. AEP structure is as illustrated in Figure 3.2. Diverse maximum normalized algorithms including MNLMS, MENLMS, MNLMF, MVNLMS and their sign-based versions are used to implement multiple AEPs. To compare, LMS-based AEP is also implemented. We acquired ten DNA sequences for assessment from the NCBI nucleotide gene databank [40]. For consistent outcomes, we considered these sequences to assess the efficiency of multiple algorithms. The description of several gene sequences taken from the NCBI gene databank is shown in Table 3.8.

4.10.1 Exon Prediction Results

Here, the performance measures of various AEPs developed using maximum normalized based adaptive algorithms are derived, analyzed, and compared. Various maximum normalized based AEPs are developed using MNLMS, MENLMS, MNLMF, and MVNLMS algorithms along with their sign-based variants. To evaluate the efficiency of various developed AEPs, we have considered exon prediction results for ten gene sequences from the NCBI gene database. The performance metrics for sequence 5 with accession AF009962 for all NLMS and MNLMS algorithms and their signed variants are measured using MATLAB as shown in Tables 4.7–4.10.

The performance measures Sn, Sp, and Pr are assessed at threshold values from 0.4 to 0.9 with an interval of 0.05. The exon identification at threshold 0.8 appears to be better, and the values of performance measures using NLMS- and MNLMS-based algorithms along with their signed versions are as shown in Table 4.7. From these values, it is evident that the performance measures of NLMS- and MNLMS-based AEPs are just inferior compared to NSRLMS- and MNSRLMS-based AEPs. For a larger number of iterations, the computational complexity of NLMS and MNLMS algorithms increases than NSRLMS and MNSRLMS algorithms. In such cases, the MNSRLMS algorithm becomes efficient due to its significantly low complexity in performing the computations and in terms of exon locating ability.

The signum function present in all signed versions of MNLMS reduces the computational complexity, and thus, all signed versions predict the exon locations more accurately. Of all these algorithms, MNSRLMS-based AEP is effective in terms of accurate exon prediction when compared to LMS, NLMS, and its other signed variants with sensitivity Sn 0.6954 (69.54%), specificity Sp 0.7011 (70.11%), and precision Pr 0.6567 (65.67%) respectively, which are a bit inferior than MNLMS-based AEP. But because of its significantly low computational complexity, MNSRLMS-based AEP outperforms all other NLMS- and MNLMS-based adaptive variants. A typical PSD plot for MNLMS and its sign-based variants is shown in Figure 4.10. It shows

TABLE 4.7
Performance Measures of NLMS- and MNLMS-Based AEPs for a Gene Sequence with Accession AF009962

		Gene Sequence Serial Number									
Algorithm	**Metric**	**1**	**2**	**3**	**4**	**5**	**6**	**7**	**8**	**9**	**10**
LMS	Sn	0.6286	0.6384	0.6457	0.6273	0.6481	0.6162	0.6193	0.6241	0.6268	0.6202
	Sp	0.6435	0.6628	0.6587	0.6405	0.6518	0.6324	0.6529	0.6289	0.6452	0.5965
	Pr	0.5922	0.5894	0.5934	0.5858	0.5904	0.5786	0.5896	0.5856	0.5814	0.5761
NLMS	Sn	0.7085	0.7137	0.7227	0.7086	0.7240	0.7162	0.7192	0.7162	0.7285	0.7286
	Sp	0.7267	0.7458	0.7321	0.7278	0.7378	0.7284	0.7396	0.7284	0.7393	0.6976
	Pr	0.6954	0.7027	0.6987	0.7096	0.6927	0.6857	0.6904	0.6857	0.6896	0.6825
NSRLMS	Sn	0.6872	0.6996	0.7027	0.6862	0.7026	0.6814	0.6894	0.6814	0.7074	0.7046
	Sp	0.7043	0.7263	0.7137	0.7036	0.7114	0.7035	0.7112	0.7035	0.7068	0.6738
	Pr	0.6722	0.6638	0.6602	0.6734	0.6672	0.6526	0.6793	0.6526	0.6554	0.6688
NSLMS	Sn	0.6687	0.6741	0.6838	0.6694	0.6852	0.6524	0.6602	0.6524	0.6786	0.6772
	Sp	0.6802	0.7057	0.6946	0.6851	0.6902	0.6727	0.6994	0.6727	0.6874	0.6546
	Pr	0.6545	0.6336	0.6435	0.6553	0.6433	0.6295	0.6484	0.6295	0.6285	0.6458
NSSLMS	Sn	0.6436	0.6582	0.6612	0.6462	0.6614	0.6372	0.6492	0.6372	0.6512	0.6598
	Sp	0.6642	0.6876	0.6736	0.6674	0.6707	0.6582	0.6776	0.6582	0.6642	0.6353
	Pr	0.6115	0.6195	0.6295	0.6156	0.6225	0.6084	0.6175	0.6084	0.5893	0.6156
MNLMS	Sn	0.7018	0.7013	0.7122	0.7026	0.7124	0.7016	0.7019	0.7043	0.7128	0.7202
	Sp	0.7126	0.7450	0.7232	0.7127	0.7237	0.7128	0.7239	0.7158	0.7239	0.6897
	Pr	0.6895	0.7002	0.6928	0.7049	0.6872	0.6785	0.6850	0.6785	0.6821	0.6782
MNSRLMS	Sn	0.6787	0.6899	0.7002	0.6786	0.6954	0.6781	0.6785	0.6781	0.7007	0.6978
	Sp	0.7004	0.7126	0.7013	0.6953	0.7011	0.6973	0.7011	0.6973	0.6956	0.6673
	Pr	0.6672	0.6563	0.6560	0.6673	0.6567	0.6452	0.6719	0.6452	0.6455	0.6564
MNSLMS	Sn	0.6564	0.6674	0.6783	0.6569	0.6785	0.6452	0.6560	0.6452	0.6678	0.6677
	Sp	0.6780	0.6975	0.6894	0.6785	0.6892	0.6672	0.6879	0.6672	0.6787	0.6452
	Pr	0.6454	0.6233	0.6343	0.6465	0.6343	0.6169	0.6348	0.6169	0.6135	0.6345
MNSSLMS	Sn	0.6343	0.6458	0.6561	0.6346	0.6561	0.6257	0.6349	0.6257	0.6451	0.6459
	Sp	0.6564	0.6787	0.6673	0.6567	0.6670	0.6458	0.6677	0.6458	0.6564	0.6266
	Pr	0.6011	0.6059	0.6159	0.6015	0.6122	0.6025	0.6067	0.6008	0.5789	0.6043

the predicted exon location of sequence 5 applying LMS, MNLMS, and its sign-based adaptive algorithms.

From these plots, it is evident that LMS-based AEP did not correctly predict the coding regions. This algorithm creates some ambiguities in location identification in the form of a short peak for exon position and few unwanted peaks in various power spectrum positions. In the case of MNLMS-based signed versions, MNSRLMS algorithm exactly predicted the exon location at 3934-4581 with good intensity and a sharp peak in PSD plot is observed.

The PSD plots for MNLMS and its signed variants are shown in Figure 4.10b–e. Because of the sign function in these algorithms, these algorithms are better in terms of tracking ability of exons than LMS algorithm. With MNSSLMS, the efficiency of

TABLE 4.8
Performance Measures of ENLMS- and MENLMS-Based AEPs for a Gene Sequence with Accession AF009962

		Gene Sequence Serial Number									
Algorithm	**Metric**	**1**	**2**	**3**	**4**	**5**	**6**	**7**	**8**	**9**	**10**
LMS	Sn	0.6286	0.6384	0.6457	0.6273	0.6481	0.6162	0.6193	0.6241	0.6268	0.6202
	Sp	0.6435	0.6628	0.6587	0.6405	0.6518	0.6324	0.6529	0.6289	0.6452	0.5965
	Pr	0.5922	0.5894	0.5934	0.5858	0.5904	0.5786	0.5896	0.5856	0.5814	0.5761
ENLMS	Sn	0.7027	0.6936	0.7031	0.7046	0.6992	0.7055	0.7036	0.7098	0.7031	0.7098
	Sp	0.7126	0.7242	0.7242	0.7125	0.7137	0.7198	0.7224	0.7061	0.7222	0.6836
	Pr	0.6812	0.6845	0.6755	0.6741	0.6734	0.6726	0.6743	0.6744	0.6735	0.6726
ENSRLMS	Sn	0.6834	0.6751	0.6884	0.6846	0.6858	0.6905	0.6837	0.6838	0.6838	0.6866
	Sp	0.6935	0.7102	0.7012	0.6935	0.6933	0.6929	0.7032	0.6922	0.6993	0.6534
	Pr	0.6543	0.6572	0.6536	0.6586	0.6457	0.6566	0.6563	0.6547	0.6582	0.6573
ENSLMS	Sn	0.6612	0.6504	0.6648	0.6693	0.6652	0.6635	0.6635	0.6654	0.6656	0.6656
	Sp	0.6879	0.6961	0.6827	0.6722	0.6677	0.6664	0.6852	0.6899	0.6817	0.6344
	Pr	0.6301	0.6323	0.6339	0.6324	0.6252	0.6327	0.6347	0.6336	0.6334	0.6325
ENSSLMS	Sn	0.6332	0.6376	0.6368	0.6412	0.6380	0.6479	0.6436	0.6442	0.6472	0.6434
	Sp	0.6586	0.6745	0.6705	0.6582	0.6574	0.6486	0.6625	0.6355	0.6936	0.6220
	Pr	0.6156	0.6044	0.6131	0.6135	0.6045	0.6142	0.6141	0.6193	0.6191	0.6184
MENLMS	Sn	0.7002	0.6893	0.7003	0.6994	0.6889	0.7005	0.7013	0.7029	0.7003	0.7031
	Sp	0.7012	0.7124	0.7124	0.7012	0.7013	0.7019	0.7112	0.7016	0.7124	0.6783
	Pr	0.6781	0.6784	0.6675	0.6674	0.6673	0.6672	0.6674	0.6674	0.6673	0.6672
MENSRLMS	Sn	0.6783	0.6675	0.6788	0.6784	0.6785	0.6890	0.6783	0.6783	0.6783	0.6786
	Sp	0.6893	0.7010	0.7001	0.6893	0.6893	0.6892	0.7003	0.6892	0.6889	0.6453
	Pr	0.6454	0.6457	0.6432	0.6458	0.6348	0.6456	0.6456	0.6454	0.6458	0.6457
MENSLMS	Sn	0.6561	0.6450	0.6564	0.6569	0.6565	0.6563	0.6563	0.6565	0.6566	0.6565
	Sp	0.6787	0.6896	0.6782	0.6672	0.6577	0.6566	0.6785	0.6789	0.6781	0.6254
	Pr	0.6230	0.6232	0.6233	0.6232	0.6125	0.6232	0.6234	0.6233	0.6233	0.6272
MENSSLMS	Sn	0.6233	0.6237	0.6236	0.6341	0.6238	0.6347	0.6343	0.6344	0.6347	0.6363
	Sp	0.6458	0.6674	0.6670	0.6458	0.6457	0.6348	0.6562	0.6235	0.6873	0.6122
	Pr	0.6075	0.6004	0.6013	0.6013	0.6004	0.6015	0.6112	0.6019	0.6122	0.6135

exon prediction is inferior to its other signed variants because of the clipped input sequence and clipped feedback signal. Based on less computational complexity, performance measures in Table 4.7, and exon prediction plots in Figure 4.10, it is found that MNSRLMS-based AEP is found to be a better candidate in exon prediction applications.

In Figure 4.10a, a relatively short peak is evident for exon location, and some undesirable peaks are evident at different locations in the power spectrum plot using LMS algorithm. At the same moment, LMS algorithm has not correctly identified the actual exon position at 3934-4581. It was clear that MNSRLMS-based AEP predicted the exon location with a sharp peak accurately compared to all other MNLMS variants. The values of prediction measures such as sensitivity, specificity, and precision

TABLE 4.9
Performance Measures of NLMF- and MNLMF-Based AEPs for a Gene Sequence with Accession AF009962

		Gene Sequence Serial Number									
Algorithm	**Metric**	**1**	**2**	**3**	**4**	**5**	**6**	**7**	**8**	**9**	**10**
LMS	Sn	0.6286	0.6384	0.6457	0.6273	0.6481	0.6162	0.6193	0.6241	0.6268	0.6202
	Sp	0.6435	0.6628	0.6587	0.6405	0.6518	0.6324	0.6529	0.6289	0.6452	0.5965
	Pr	0.5922	0.5894	0.5934	0.5858	0.5904	0.5786	0.5896	0.5856	0.5814	0.5761
NLMF	Sn	0.7486	0.7334	0.7293	0.7365	0.7298	0.7326	0.7346	0.7363	0.7382	0.7393
	Sp	0.7375	0.7435	0.7512	0.7471	0.7432	0.7432	0.7478	0.7414	0.7453	0.7076
	Pr	0.7352	0.7354	0.7226	0.7287	0.6895	0.7154	0.7328	0.7206	0.7236	0.7267
NSRLMF	Sn	0.7348	0.7233	0.7129	0.7236	0.7129	0.7232	0.7254	0.7236	0.7238	0.7239
	Sp	0.7237	0.7326	0.7451	0.7347	0.7343	0.7343	0.7347	0.7341	0.7345	0.7007
	Pr	0.7255	0.7246	0.7122	0.7128	0.6789	0.7015	0.7232	0.7120	0.7113	0.7106
NSLMF	Sn	0.7234	0.7163	0.7052	0.7153	0.7042	0.7173	0.7125	0.7153	0.7125	0.7193
	Sp	0.7143	0.7232	0.7345	0.7234	0.7234	0.7284	0.7262	0.7234	0.7251	0.6986
	Pr	0.7185	0.7124	0.7104	0.7052	0.6708	0.7001	0.7173	0.7012	0.7031	0.7010
NSSLMF	Sn	0.7205	0.7118	0.7005	0.7015	0.7004	0.7017	0.7012	0.7115	0.7013	0.7129
	Sp	0.7101	0.7133	0.7214	0.7152	0.7153	0.7128	0.7126	0.7163	0.7135	0.6898
	Pr	0.7088	0.7032	0.7010	0.7002	0.6670	0.6972	0.7017	0.7001	0.6993	0.6986
MNLMF	Sn	0.7324	0.7218	0.7187	0.7259	0.7192	0.7117	0.7237	0.7237	0.7267	0.7277
	Sp	0.7287	0.7387	0.7415	0.7314	0.7321	0.7367	0.7362	0.7306	0.7345	0.6986
	Pr	0.7230	0.7254	0.7113	0.7172	0.6758	0.7069	0.7287	0.7186	0.7127	0.7132
MNSRLMF	Sn	0.7257	0.7152	0.7027	0.7176	0.7018	0.7142	0.7162	0.7184	0.7127	0.7154
	Sp	0.7124	0.7268	0.7342	0.7258	0.7278	0.7285	0.7224	0.7234	0.7262	0.6913
	Pr	0.7163	0.7107	0.7018	0.7035	0.6869	0.6907	0.7152	0.7004	0.7049	0.7029
MNSLMF	Sn	0.7147	0.7042	0.6978	0.7012	0.6926	0.7037	0.7056	0.7086	0.7045	0.7059
	Sp	0.7076	0.7176	0.7282	0.7187	0.7154	0.7174	0.7145	0.7115	0.7137	0.6866
	Pr	0.7037	0.7027	0.7018	0.6934	0.6617	0.6911	0.7025	0.6928	0.6941	0.6939
MNSSLMF	Sn	0.7116	0.7043	0.6913	0.6956	0.6915	0.6928	0.6972	0.7046	0.6949	0.7018
	Sp	0.7009	0.7072	0.7167	0.7022	0.7021	0.7010	0.7019	0.7072	0.7031	0.6726
	Pr	0.6976	0.6943	0.6921	0.6947	0.5681	0.6825	0.6923	0.6951	0.6877	0.6833

of SRLMS, SLMS, and SSLMS algorithms from Table 3.8 are observed as inferior to all MNLMS-based adaptive algorithms.

Similarly, the performance measures of sequence 5 with accession AF009962 for all ENLMS- and MENLMS-based algorithms along with their signed versions are as shown in Table 4.8. From these values, it is evident that the performance measures of ENLMS- and MENLMS-based AEPs are just inferior compared to ENSRLMS- and MENSRLMS-based AEPs. For a larger number of iterations, the computational complexity of ENLMS and MENLMS algorithms is higher than ENSRLMS and MENSRLMS algorithms. In such cases, the MENSRLMS algorithm becomes efficient due to its significantly low complexity in performing the computations and in terms of exon locating ability.

TABLE 4.10
Performance Measures of MVNLMS- and MVNSRLMS-Based AEPs for a Gene Sequence with Accession AF009962

		Gene Sequence Serial Number									
Algorithm	**Metric**	**1**	**2**	**3**	**4**	**5**	**6**	**7**	**8**	**9**	**10**
LMS	Sn	0.6286	0.6384	0.6457	0.6273	0.6481	0.6162	0.6193	0.6241	0.6268	0.6202
	Sp	0.6435	0.6628	0.6587	0.6405	0.6518	0.6324	0.6529	0.6289	0.6452	0.5965
	Pr	0.5922	0.5894	0.5934	0.5858	0.5904	0.5786	0.5896	0.5856	0.5814	0.5761
VNLMS	Sn	0.8128	0.8024	0.8028	0.8145	0.7989	0.8132	0.8062	0.8010	0.8190	0.7946
	Sp	0.8021	0.7992	0.8121	0.8024	0.8058	0.8060	0.7917	0.8173	0.8056	0.8072
	Pr	0.8137	0.8136	0.7994	0.8137	0.8121	0.8122	0.8198	0.7957	0.8109	0.8157
VNSRLMS	Sn	0.7972	0.7835	0.7882	0.7936	0.7884	0.7978	0.7873	0.7868	0.7998	0.7834
	Sp	0.7836	0.7841	0.7936	0.7835	0.7812	0.7828	0.7851	0.7997	0.7884	0.7837
	Pr	0.7783	0.7924	0.7823	0.7941	0.7936	0.7712	0.7966	0.7848	0.7959	0.7987
VNSLMS	Sn	0.7795	0.7769	0.7793	0.7735	0.7615	0.7737	0.7715	0.7767	0.7754	0.7699
	Sp	0.7732	0.7685	0.7592	0.7529	0.7596	0.7726	0.7633	0.7544	0.7548	0.7512
	Pr	0.7697	0.7715	0.7667	0.7775	0.7782	0.7673	0.7760	0.7698	0.7717	0.7749
VNSSLMS	Sn	0.7581	0.7597	0.7581	0.7557	0.7526	0.7576	0.7571	0.7549	0.7537	0.7539
	Sp	0.7565	0.7586	0.7465	0.7497	0.7461	0.7589	0.7541	0.7437	0.7486	0.7455
	Pr	0.7488	0.7526	0.7488	0.7586	0.7565	0.7447	0.7585	0.7428	0.7512	0.7577
MVNLMS	Sn	0.7692	0.7691	0.7692	0.7638	0.7663	0.7611	0.7601	0.7676	0.7653	0.7612
	Sp	0.7684	0.7635	0.7682	0.7691	0.7692	0.7612	0.7619	0.7610	0.7673	0.7659
	Pr	0.7595	0.7463	0.7596	0.7598	0.7682	0.7576	0.7457	0.7544	0.7567	0.7644
MVNSRLMS	Sn	0.7507	0.7456	0.7517	0.7537	0.7546	0.7525	0.7411	0.7537	0.7514	0.7569
	Sp	0.7423	0.7523	0.7423	0.7476	0.7583	0.7455	0.7537	0.7469	0.7433	0.7575
	Pr	0.7512	0.7392	0.7532	0.7443	0.7525	0.7530	0.7374	0.7558	0.7467	0.7580
MVNSLMS	Sn	0.7416	0.7432	0.7446	0.7374	0.7457	0.7477	0.7490	0.7408	0.7399	0.7464
	Sp	0.7465	0.7476	0.7365	0.7402	0.7446	0.7418	0.7439	0.7349	0.7427	0.7433
	Pr	0.7396	0.7257	0.7456	0.7376	0.7465	0.7324	0.7233	0.7466	0.7369	0.7490
MVNSSLMS	Sn	0.7302	0.7318	0.7306	0.7214	0.7306	0.7398	0.7318	0.7306	0.7214	0.7306
	Sp	0.7212	0.7311	0.7212	0.7318	0.7382	0.7273	0.7311	0.7212	0.7318	0.7382
	Pr	0.7323	0.7186	0.7323	0.7251	0.7296	0.7358	0.7186	0.7323	0.7251	0.7296

The signum function present in all signed versions of MENLMS reduces the computational complexity, and they have predicted the exon locations more accurately compared to LMS. Of all these algorithms, MENSRLMS-based AEP is effective in terms of accurate exon prediction when compared to LMS, ENLMS, and its other signed variants with sensitivity Sn 0.6785 (67.85%), specificity Sp 0.6893 (68.93%), and precision Pr 0.6348 (63.48%) respectively, which are just inferior than MENLMS-based AEP. But, because of its significantly low computational complexity, MENSRLMS-based AEP outperforms all other ENLMS- and MENLMS-based adaptive variants.

At a threshold of 0.8, the exon prediction seems better for sign regressor–based AEP for a gene sequence with accession AF009962. A typical PSD plot for

MENLMS and its sign-based variants is shown in Figure 4.10. It shows the predicted exon location of sequence 5 applying LMS, MENLMS, and its sign-based adaptive algorithms. From these plots, it is obvious that coding regions are not correctly predicted for AEP based on LMS. This algorithm generates certain ambiguities in location prediction in the form of a short peak for exon location and some unwanted peaks at different positions in the power spectrum. In the case of MENLMS-based signed versions, MENSRLMS-based AEP exactly predicted the exon location at 3934-4581 with great intensity and a sharp peak in PSD plot is observed. Due to use of signum function in MENSRLMS-based AEP, it offers good exon location ability and low computational complexity compared to LMS, MENLMS, and its other signed variants. The PSD plots for MENLMS and its signed variants are depicted in Figure 4.11b–e.

Due to the sign function of these algorithms, their tracking ability of exons stands better than LMS algorithm. With MENSSLMS, the efficiency of the exon prediction is lower than other signed variants because of clipped input sequence and clipped feedback signal. It is therefore deliberated that MENSRLMS-based AEP is the best choice in exon identification applications on the basis of low computational difficulty, performance measurements in Table 4.8, and exon prediction plots in Figure 4.11. In Figure 4.11a, a relatively short peak is evident for exon location and certain unwanted

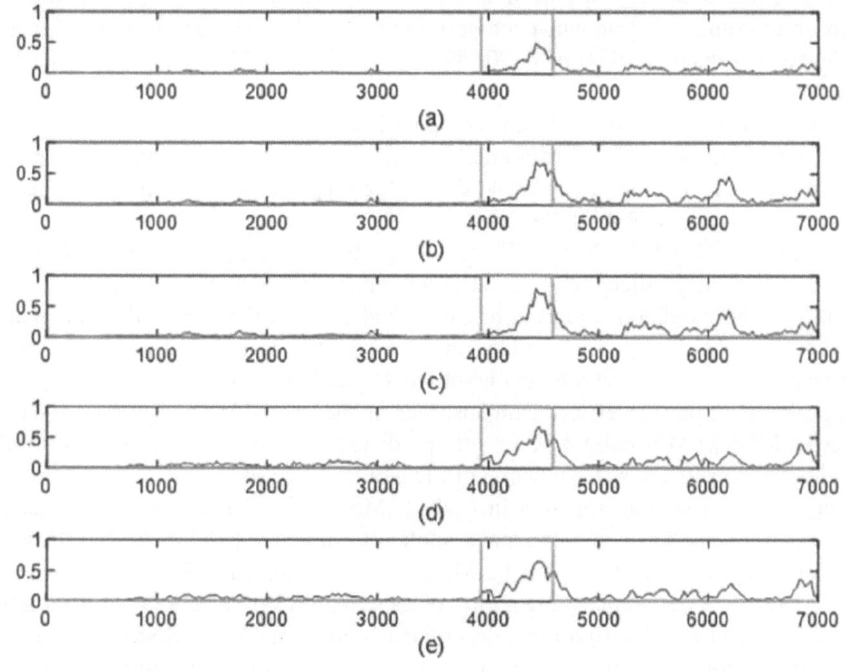

FIGURE 4.11 PSD plots with location of exon (3934-4581) for a DNA sequence with accession AF009962 predicted using (a) LMS-based AEP, (b) MENLMS-based AEP, (c) MENSRLMS-based AEP, (d) MENSLMS-based AEP, and (e) MENSSLMS-based AEP. (Relative base location is taken on *x*-axis, and power spectrum is taken on *y*-axis.)

peaks are found at different locations in the power spectrum plot using LMS algorithm. At the same moment, LMS algorithm does not correctly locate the actual exon location at 3934-4581. It is clear that MENSRLMS-based AEP predicted the exon location with a sharp peak accurately compared to all other MNLMS variants.

The values of prediction measures such as sensitivity, specificity, and precision of SRLMS, SLMS, and SSLMS algorithms from Table 3.8 are observed as inferior to all MENLMS-based adaptive algorithms. Due to use of signum function in MENSRLMS-based AEP, it offers good exon location ability and low computational complexity compared to LMS, MENLMS, and its other signed variants. Likewise, the performance measures of sequence 5 with accession AF009962 for all NLMF- and MNLMF-based algorithms along with their signed versions are as shown in Table 4.9. From these values, it is evident that the performance measures of NLMF- and MNLMF-based AEPs are just inferior compared to NSRLMF- and MNSRLMF-based AEPs. For a larger number of iterations, the computational complexity of NLMF and MNLMF algorithms is higher than NSRLMF and MNSRLMF algorithms.

In such cases, the MNSRLMF algorithm becomes efficient due to its significantly low complexity in performing the computations and in terms of exon locating ability. The signum function present in all signed versions of MNLMF reduces the computational complexity, and they have predicted the exon locations more accurately compared to LMS. Of all these algorithms, MNSRLMF-based AEP is effective in terms of accurate exon prediction when compared to LMS, MNLMF, and its other signed variants with sensitivity Sn 0.7018 (70.18%), specificity Sp 0.7278 (72.78%), and precision Pr 0.6869 (68.69%) respectively, which are just inferior than MNLMF-based AEP. But, due to its significantly low computational complexity, MNSRLMF-based AEP outperforms all other NLMF- and MNLMF-based adaptive variants. Exon prediction seems to be better at threshold 0.8 for sign regressor–based AEP for a gene sequence with accession AF009962.

A typical PSD plot for MNLMF and its sign-based variants is shown in Figure 4.12. It shows the predicted exon location of sequence 5 applying LMS, MNLMF, and its sign-based adaptive algorithms. It is evident from these plots that the exon position has not correctly predicted by LMS-based AEP. This algorithm creates certain ambiguities in location prediction with an evident short peak and some undesirable peaks at various power spectrum positions. In the case of MNLMF-based signed versions, MNSRLMF-based AEP exactly predicted the exon location at 3934-4581 with good intensity and a sharp peak in PSD plot is observed.

Due to use of signum function in MNSRLMF-based AEP, it offers good exon location ability and low computational complexity compared to LMS, MNLMF, and its other signed variants. PSD plots for MNLMF and its signed variants are shown in Figure 4.12b–e. The exon locating ability of these algorithms is better than the LMS algorithm because of the sign function in those algorithms. With MNSSLMF algorithm, the performance of the exon prediction is less than its other signed variants because of the clipped input sequence and the clipped feedback signal. Therefore, it is shown that MNSRLMF-based AEP is found to be a better candidate in exon prediction applications on the basis of low computational complexity, performance measures in Table 4.9, and exon prediction plots in Figure 4.12.

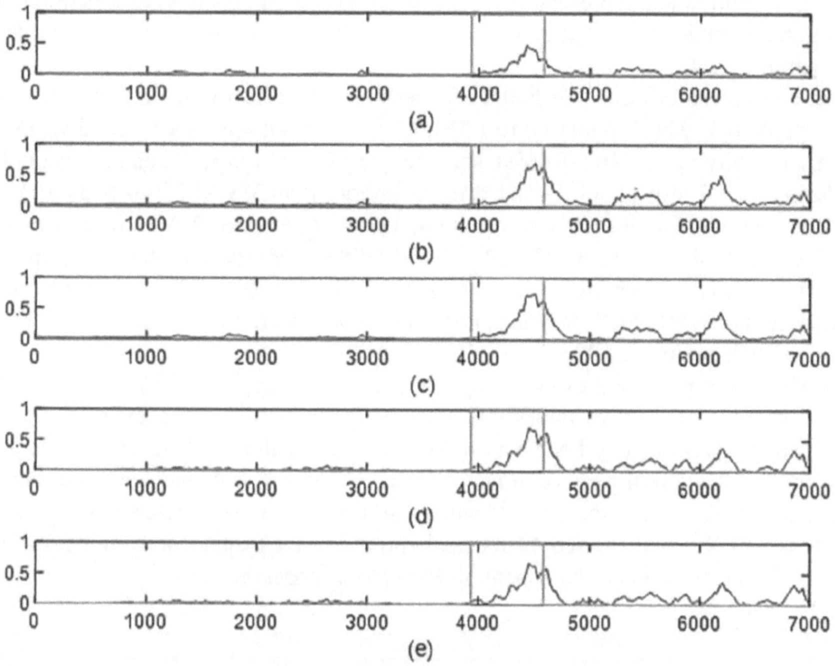

FIGURE 4.12 PSD plots with the location of exon (3934-4581) for a DNA sequence with accession AF009962 predicted using (a) LMS-based AEP, (b) MNLMF-based AEP, (c) MNSRLMF-based AEP, (d) MNSLMF-based AEP, and (e) MNSSLMF-based AEP. (Relative base location is taken on *x*-axis, and power spectrum is taken on *y*-axis.)

In Figure 4.12a, a relatively short peak is evident for exon location and certain unwanted peaks are identified at different locations in PSD plot using LMS algorithm. At the same moment, LMS algorithm does not locate the actual exon position at 3934-4581. It is clear that MNSRLMF-based AEP predicted the exon location with a sharp peak accurately compared to all other MNLMF variants. The values of prediction measures such as sensitivity, specificity, and precision of SRLMS, SLMS, and SSLMS algorithms from Table 3.4 are observed as inferior to all MNLMF-based adaptive algorithms. Due to use of signum function in MNSRLMF-based AEP, it offers good exon location ability and low computational complexity compared to LMS, MNLMF, and its other signed variants.

Also, the performance measures of sequence 5 with accession AF009962 for all VNLMS- and MVNLMS-based algorithms along with their signed versions are as shown in Table 4.10. From these values, it is evident that the performance measures of VNLMS- and MVNLMS-based AEPs are just inferior compared to VNSRLMS- and MVNSRLMS-based AEPs. For a larger number of iterations, the computational complexity of VNLMS and MVNLMS algorithms is higher than MVNSRLMS and MVNSRLMS algorithms. In such cases, the MVNSRLMS algorithm becomes efficient due to its significantly low complexity in performing the computations and in terms of exon locating ability.

The signum function present in all signed versions of MVNLMS reduces the computational complexity, and they have predicted the exon locations more accurately compared to LMS.

Of all these algorithms, MVNSRLMS-based AEP is effective in terms of accurate exon prediction when compared to LMS, MVNLMS, and its other signed variants with sensitivity Sn 0.7546 (75.46%), specificity Sp 0.7583 (75.83%), and precision Pr 0.7525 (75.25%) respectively, which are just inferior than MVNLMS-based AEP.

But, due to its significantly low computational complexity, MVNSRLMS-based AEP outperforms all other VNLMS- and MVNSRLMS-based adaptive variants. Exon prediction seems better at threshold 0.8 for SRA-based AEP for a gene sequence with accession AF009962. A typical PSD plot for MVNLMS and its sign-based variants is shown in Figure 4.13.

It shows the predicted exon location of sequence 5 applying LMS, MVNLMS, and its sign-based adaptive algorithms. It is true from these plots that exon position is not correctly predicted by LMS-based AEP. This algorithm leads to some ambiguities in identification of exon sections in terms of a small peak and fewer unwanted peaks at distinct power spectrum locations. In the case of MVNLMS-based signed versions, MVNSRLMS-based AEP exactly predicted the exon location at 3934-4581 with good intensity and a sharp peak in PSD plot is observed.

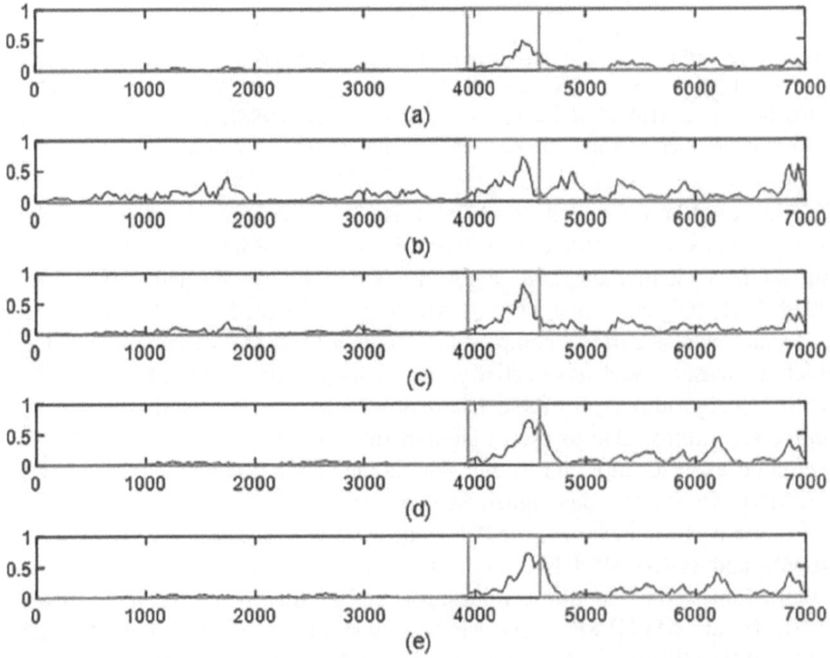

FIGURE 4.13 PSD plots with location of exon (3934-4581) for a DNA sequence with accession AF009962 predicted using (a) LMS-based AEP, (b) MVNLMS-based AEP, (c) MVNSRLMS-based AEP, (d) MVNSLMS-based AEP, and (e) MVNSSLMS-based AEP. (Relative base location is taken on x-axis, and power spectrum is taken on y-axis.)

Due to use of signum function in MVNSRLMS-based AEP, it offers good exon location ability and low computational complexity compared to LMS, MVNLMS, and its other signed variants. The PSD plots and their signed variants for MVNLMS are displayed in Figure 4.13b–e. The exon locating capability of these algorithms is better than LMS algorithm due to sign function within these algorithms. The output of exon prediction is lower compared to its other signed variants because of clipped input sequence and clipped feedback signal in the case of MVNSSLMS technique. Consequently, AEP based on MVNSRLMS is deliberated to be a better candidate in exon locating applications based on lesser computational complexity, performance measures presented in Table 4.10, and exon prediction plots shown in Figure 4.13. In Figure 4.13a, a relatively short peak is evident for exon location and certain unwanted peaks are identified at different locations in the power spectrum plot using LMS algorithm.

At the same moment, LMS algorithm has not located the real exon position at 3934-4581. It was clear that MVNSRLMS-based AEP predicted the exon location with a sharp peak accurately compared to all other MVNLMS variants. The values of prediction measures such as sensitivity, specificity, and precision of SRLMS, SLMS, and SSLMS algorithms from Table 3.9 are observed as inferior to all MVNLMS-based adaptive algorithms. Due to use of signum function in MVNSRLMS-based AEP, it offers good exon location ability and low computational complexity compared to LMS, MVNLMS, and its other signed variants.

4.11 CONCLUSIONS

In this chapter, we have implemented distinct weight update versions of normalization-based realization of various adaptive algorithms. These algorithms are used to develop various AEPs for accurate prediction of exon locations in real gene sequences of *Homo sapiens*. Varieties of AEPs are examined with sixteen adaptive algorithms based on normalization with their sign-based variants on original DNA datasets of *Homo sapiens* taken from the NCBI gene databank.

An introduction to normalized adaptive algorithms, an account of detailed explanation about different normalized adaptive algorithms including NLMS, ENLMS, NMLF, and VNLMS, and an extension to their sign-based realizations are presented. The LMS algorithm is considered as the reference algorithm, and various normalization-based realizations are presented. In order to evaluate the efficiency of different AEPs, we have calculated convergence features; PSD plots also computational complexities, as well as measures such as sensitivity (Sn), specificity (Sp), and precision (Pr). We have inspected the possibility of enhancing the existing methods in order to achieve a better performance.

To achieve this, reducing computational difficulty and improving exon tracking capability are desirable. This can be accomplished by either clipping input or estimation error. Based on this clipping, three signed algorithms are obtained. We have extended the NLMS-based adaptive algorithms to three versions of sign-based adaptive algorithms. To further reduce the computational complexity significantly and provide a better performance, we have extended all the normalized algorithms to maximum normalized based adaptive algorithms. This is the second category of the adaptive algorithms we have considered. Totally, thirty-two adaptive exon predictors are developed and tested on ten real gene sequences of *Homo sapiens* taken from the NCBI databank. Of these, sixteen normalized adaptive algorithms along their sign-based algorithms are used for development of various AEPs including NLMS, NSRLMS, NSLMS, NSSLMS, ENLMS, ENSRLMS, ENSLMS, ENSSLMS, NLMF, NSRLMF, NSLMF, NSSLMF, VNLMS, VNSRLMS, VNSLMS, and VNSSLMS algorithms. Among all the 16 normalized adaptive algorithms, VNSRLMS-based AEP performed better in terms of exon tracking ability, low computational complexity, and convergence performance.

Similarly, various AEPs are developed the sixteen maximum normalized adaptive algorithms and their sign-based algorithms including MNLMS, MNSRLMS, MNSLMS, MNSSLMS, MENLMS, MENSRLMS, MENSLMS, MENSSLMS, MNLMF, MNSRLMF, MNSLMF, MNSSLMF, MVNLMS, MVNSRLMS, MVNSLMS, and MVNSSLMS algorithms. Among all these 16 maximum normalized adaptive algorithms, MVNSRLMS-based AEP performed better in terms of exon tracking ability, low computational complexity, and convergence performance.

Between the five normalized based algorithms, VNLMS algorithm performs better compared to all other normalized adaptive algorithms in terms of exon tracking ability and low computational complexity. Among all the sign-based normalized and maximum normalized adaptive algorithms, all their sign regressor versions performed better. All their SA- and SSA-based AEPs are slightly inferior to their LMS counterpart. Overall all their proposed sign LMS-based AEPs deliver a greater performance than the existing LMS pertaining to computational complexity and metrics like sensitivity, precision, and specificity attained using a gene sequence with accession AF009962 at a threshold value of 0.8. Based on the assessment of multiple AEPs addressed in this chapter, MVNSRLMS-based AEP is found to be better in terms of exon forecast accuracy among all normalization and maximum normalization-based AEPs, performance measures, and convergence performance due to its low computational complexity. Henceforth, MVNSRLMS adaptive algorithm is more appropriate for use in system-on-chip (SOC)- and lab-on-a-chip (LOC)-based applications to develop low-complexity nano-bioinformatics devices.

5 Logarithmic-Based Realization of Adaptive Filtering Techniques for Exon Prediction

5.1 INTRODUCTION

In the field of bioinformatics, precisely locating the exon segments in a DNA sequence based on three base periodicity (TBP) has a vibrant effect on disease diagnosis and medication design. TBP property has been displayed with most of the DNA sequences. Numerous existing methods using techniques relevant to signal processing for locating gene locations in a DNA sequence are proved less efficient. In order to achieve fast convergence and good filtering capability, we introduce normalized logarithmic realization of adaptive filtering algorithms and their sign-based versions to reduce computational burden in this chapter.

Adaptive algorithms are used for processing of genomic signals of long length which can change weight coefficients depending on cost function. Least mean square (LMS) is a basic, easy-to-execute adaptive technique. Input data vector is proportionate to the weight update process in LMS, and also its fixed step size is determined. Therefore, when data input vector is big, LMS algorithm suffers from setbacks such as drift weight, gradient noise amplification, and poor convergence. In applications of adaptive filtering such as channel equalization and noise suppression, a certain statistical error of $e(n)$ indicating the difference between desired signal $d(n)$ and estimated response is used. Usually, owing to mathematical tractability and comparative ease of evaluation, a cost function MSE is used. The members of this class are LMS and normalized LMS (NLMS) algorithms. We have used logarithmic adaptive techniques dependent on relative logarithmic cost function to overcome these LMS constraints.

Inherently, logarithmic algorithms combine greater- and lower-order measurements into single error update based on the amount of error. The resulting logarithmic adaptive algorithms include least mean logarithmic square (LMLS) and least logarithmic absolute difference (LLAD) algorithms, which enhance convergence efficiency and decrease the computational complexity of conventional LMS algorithm [124]. LMLS algorithm achieves similar convergence efficiency and expands its stability bound at the step size with LMF algorithm. The LLAD and LMS implementations demonstrate similar convergence effectiveness in the impulse-free noise environments, whereas LLAD is strong against impulsive interferences and passes SA.

As stated earlier in Chapter 3, LMS has two significant disadvantages. Obviously, the data input vector directly relational to fixed step size and weight update mechanism of LMS algorithm are determined. Therefore, when the data vector is larger, LMS remains affected by a problem with gradient noise amplification. The curve fitting of LMS algorithm for adaptation is presented in Ref. [125]. Normalization should therefore be introduced in order to prevent this problem. Therefore, with respect to the Euclidian norm of vector input, modification of the weight vector coefficient in the filter is normalized. The step size is iteratively distinct as a consequence of normalization, and proportional to converse of expected values of total energy for coefficients of input data vector that are instantaneous.

To avoid this problem, we have extended both the logarithmic algorithms to normalized variants of LMLS and LLAD along with their sign-based algorithms. The resulting algorithms are normalized LMLS (NLMLS) and normalized LLAD (NLLAD) algorithms, respectively. Also, the signed variants of NLMLS and NLLAD are NSRLMLS, NSLMLS, NSSLMLS, NSRLLAD, NSLLAD, and NSSLLAD algorithms. This normalizes the adjustment of filter weight vector coefficient in accordance with the Euclidian squared norm of vector input in each iteration.

In real-time bioinformatics applications, various gene sequences are quite large, and hence, excessively longer-length adaptive filters may be required for processing. In such instances, implementation of LMS is more expensive and complicated. Hence, block processing of information samples is used to decrease computing difficulty of adaptive algorithm substantially in such instances. Hence, the NLMLS- and NLLAD-based logarithmic adaptive algorithms along with their signed variants are further extended to use maximum data normalization.

The resulting maximum normalization-based logarithmic adaptive algorithms include maximum NLMLS (MNLMLS) and maximum NLLAD (MNLLAD) algorithms. Both are combined with signed algorithms to further reduce complexity to perform computations resulting in MNSRLMLS, MNSLMLS, MNSSLMLS, MNSRLLAD, MNSLLAD, and MNSSLLAD algorithms. All these normalized logarithmic algorithms converge quicker compared to LMS, because a variable convergence factor is used by them to reduce instantaneous output error. The summary of various normalized and maximum-normalized logarithmic adaptive algorithms in this chapter is given in Table 5.1.

5.2 LOGARITHMIC ADAPTIVE ALGORITHMS

In this chapter, we have used normalized logarithmic adaptive techniques dependent on relative logarithmic cost function. Essentially, logarithmic algorithms combine greater- and lower-order measurements to one single continuous update. LMLS and LLAD are the resulting logarithmic adaptive algorithms, which enhance convergence and decrease computational difficulty than LMS. The normalized variants of LMLS and LLAD overcome these two drawbacks of LMS by taking into consideration signal-level variations at output of filter and normalized step parameter, thus resulting in quicker converging as well as stable algorithms. LMLS and LLAD algorithms overcome LMS constraints, offer better exon locating capability, and increase convergence performance.

TABLE 5.1
List of Algorithms Used in Chapter 5

S. No.	Acronym	Name	Advantage
1	NLMLS	Normalized least mean logarithmic square	Fast convergence, avoids gradient noise amplification problem
2	NSRLMLS	Normalized sign regressor least mean logarithmic square	Fast convergence, avoids gradient noise amplification problem, only one multiplication needed to compute the recursion
3	NSLMLS	Normalized sign least mean logarithmic square	Fast convergence, avoids gradient noise amplification problem, it requires L MACs
4	NSSLMLS	Normalized sign-sign least mean logarithmic square	Fast convergence, avoids gradient noise amplification problem, zero multiplications required, enter recursion is computed with addition with sign check operation
5	MNLMLS	Maximum normalized least mean logarithmic square	Low computational complexity, faster convergence than LMS, avoids gradient noise amplification problem
6	MNSRLMLS	Maximum normalized sign regressor least mean logarithmic square	Significantly low computational complexity, avoids gradient noise amplification problem, better stability
7	MNSLMLS	Maximum normalized sign least mean logarithmic square	Low computational complexity, avoids gradient noise amplification problem, low steady-state error and better convergence
8	MNSSLMLS	Maximum normalized sign-sign least mean logarithmic square	Fast convergence, avoids gradient noise amplification problem, less computations
9	ENLMLS	Error-normalized least mean logarithmic square	Fast convergence, minimizes signal distortion
10	ENSRLMLS	Error-normalized sign regressor least mean logarithmic square	Fast convergence, minimizes signal distortion, only one multiplication needed to compute the recursion
11	ENSLMLS	Error-normalized sign least mean logarithmic square	Fast convergence, minimizes signal distortion, it requires L MACs
12	ENSSLMLS	Error-normalized sign-sign least mean logarithmic square	Fast convergence, minimizes signal distortion, zero multiplications required, enter recursion is computed with addition with sign check operation
13	MENLMLS	Maximum error-normalized least mean logarithmic square	Low computational complexity, faster convergence than LMS, avoids gradient noise amplification problem
14	MENSRLMLS	Maximum error-normalized sign regressor least mean logarithmic square	Significantly low computational complexity, avoids gradient noise amplification problem, better stability
15	MENSLMLS	Maximum error-normalized sign least mean logarithmic square	Low computational complexity, avoids gradient noise amplification problem, low steady-state error and convergence better than LMS

(Continued)

TABLE 5.1 (*Continued*)
List of Algorithms Used in Chapter 5

S. No.	Acronym	Name	Advantage
16	MENSSLMLS	Maximum error-normalized sign-sign least mean logarithmic square	Fast convergence, avoids gradient noise amplification problem, less computations
17	NLLAD	Normalized least logarithmic absolute difference	Fast convergence, avoids gradient noise amplification problem
18	NSRLLAD	Normalized sign regressor least logarithmic absolute difference	Fast convergence, avoids gradient noise amplification problem, only one multiplication needed to compute the recursion
19	NSLLAD	Normalized sign least logarithmic absolute difference	Fast convergence, avoids gradient noise amplification problem, it requires L MACs
20	NSSLLAD	Normalized sign-sign least logarithmic absolute difference	Fast convergence, avoids gradient noise amplification problem, zero multiplications required, enter recursion is computed with addition with sign check operation
21	MNLLAD	Maximum normalized least logarithmic absolute difference	Low computational complexity, faster convergence than LMS, avoids gradient noise amplification problem
22	MNSRLLAD	Maximum normalized sign regressor least logarithmic absolute difference	Significantly low computational complexity, avoids gradient noise amplification problem, better stability
23	MNSLLAD	Maximum normalized sign least logarithmic absolute difference	Low computational complexity, avoids gradient noise amplification problem, low steady-state error and better convergence
24	MNSSLLAD	Maximum normalized sign-sign least logarithmic absolute difference	Fast convergence, avoids gradient noise amplification problem, less computations
25	ENLLAD	Error-normalized least logarithmic absolute difference	Fast convergence, avoids gradient noise amplification problem, variable step size
26	ENSRLLAD	Error-normalized sign regressor least logarithmic absolute difference	Fast convergence, avoids gradient noise amplification problem, only one multiplication needed to compute the recursion, variable step size
27	ENSLLAD	Error-normalized sign least logarithmic absolute difference	Fast convergence, avoids gradient noise amplification problem, it requires L MACs, variable step size
28	ENSSLLAD	Error-normalized sign-sign least logarithmic absolute difference	Fast convergence, avoids gradient noise amplification problem, zero multiplications required, variable step size
29	MENLLAD	Maximum error-normalized least logarithmic absolute difference	Low computational complexity, faster convergence than LMS, avoids gradient noise amplification problem

(*Continued*)

TABLE 5.1 (*Continued*)
List of Algorithms Used in Chapter 5

S. No.	Acronym	Name	Advantage
30	MENSRLLAD	Maximum error-normalized sign regressor least logarithmic absolute difference	Significantly low computational complexity, avoids gradient noise amplification problem, better stability
31	MENSLLAD	Maximum error-normalized sign least logarithmic absolute difference	Low computational complexity, avoids gradient noise amplification problem, low steady-state error and convergence better than LMS
32	MENSSLLAD	Maximum error-normalized sign-sign least logarithmic absolute difference	Fast convergence, avoids gradient noise amplification problem, less computations

The correlation between the error and reference input is normalized by a variable that corresponds to a squared norm in their normalized versions. To further improve the exon locating ability and reduce the computational complexity, we considered two types of normalizations for LMLS and LLAD, specifically data and error normalization techniques. Their resulting techniques including normalized LMLS (NLMLS) algorithm, error nonlinear LMLS or error-normalized LMLS (ENLMS) algorithm, normalized LLAD (NLLAD) algorithm, and error nonlinear LLAD or error-normalized LLAD (ENLLAD) algorithm are discussed in detail in subsequent sections.

5.3 NORMALIZED LMLS (NLMLS) ALGORITHM

NLMLS is a different group of adaptive algorithms accustomed to train coefficients of adaptive filter. One of the issues with the design and implementation of the LMS adaptive filter is to select step size. This NLMLS algorithm is considered as a unique LMS algorithm application that takes into consideration signal-level variation at the filter output and chooses a logarithmic normalized cost function that leads to a faster converging as well as stable adaptation algorithm. Normalized LMLS version based on relative logarithmic cost is referred to as the NLMLS algorithm. NLMLS algorithm overwhelms LMS limitations and enhances the speed of convergence as well as exon tracking ability (Table 5.2).

$w(n)$ is the tap-weight vector with the taps stored in a row vector

$$=\left[w(n)\ w(n-1)\ldots\ w(n-T+1)\right]^T$$

Here, an unknown vector ω_0 is presented with a linear model as

$$d(n)=\omega_o{}^T x(n)+n_t \tag{5.1}$$

where the input signal is denoted by $x(n)$ and the desired signal by $d(n)$, and n_t represents the noise.

TABLE 5.2
Mathematical Modeling of NLMLS Algorithm

Parameters: T = number of taps (i.e. filter length), μ = step size parameter

Let the tap input be $x(n)$ and filter length T is moderate to large.

Initialization: Set $w(0) = 0$ as the initial condition.

Data: Given $x(n)$ = T-by-1 tap input vector to filter $n2$ at time $n = [x(n), x(n-1) \ldots x(n-T+1)]^T$

$w(n)$ is the tap-weight vector of adaptive filter, $d(n)$ is desired response at time n, $e(n)$ is the error signal, ω_0 is an unknown vector, μ' is the adaptation constant, and $(.)^T$ is the transpose of (.).

To be computed: $w(n+1)$ = estimate of tap-weight vector at time $n+1$

Computation:

The FIR filter output is given by $y(n) = x^T(n)w(n) = w^T(n)x(n)$

The conventional cost function of the error signal $e(n)$ is $F[e(n)] = E\left[(e(n))^2\right] = E[|e(n)|]$.

Here, the normalized error cost function introduced using logarithmic function is given by

$$J(e(n)) = F(e(n)) - \frac{1}{\alpha}\ln\left(1 + \alpha F(e(n))\right) \quad (5.2)$$

where $\alpha > 0$ is a design parameter and $F(e(n))$ is a conventional cost function of error signal $e(n)$. The overall steepest descent update is achieved according to the $J(e(n))$ gradient as

$$w(n+1) = w(n) - \Delta_w \cdot F(e(n))\left[\frac{\alpha F(e(n))}{1 + \alpha F(e(n))}\right] \quad (5.3)$$

Thus, for $F(e(n)) = E(e(n))^2$, the weight update equation of the LMLS algorithm becomes

$$w(n+1) = w(n) + \mu . x(n)e(n)\left[\frac{\alpha (e(n))^2}{1 + \alpha (e(n))^2}\right] \quad (5.6)$$

Setting the time-varying size of step parameter as

$$\mu(n) = \frac{\mu'}{x^T(n)x(n)} = \frac{\mu'}{\|x(n)\|^2} \quad (5.10)$$

A small positive constant ε is added to prevent denominator from becoming too low:

$$\mu(n) = \frac{\mu'}{\varepsilon + \|x(n)\|^2} \quad (5.11)$$

The alternate weight expression for NLMLS algorithm is written as

$$w(n+1) = w(n) + \frac{\mu'}{\varepsilon + \|x(n)\|2}\, x(n)\, e(n)\left[\frac{\alpha (e(n))^2}{1 + \alpha (e(n))^2}\right]. \quad (5.14)$$

Thus, the weight update relation for maximum NLMLS for $x_{Li} \neq 0$ and $\varepsilon \neq 0$ becomes

$$w(n+1) = w(n) + \frac{\mu'}{\varepsilon + \max\left(\|x(n)\|\right)^2} x(n)e(n)\left[\frac{\alpha (e(n))^2}{1 + \alpha (e(n))^2}\right] \quad (5.17)$$

The unknown system vector here is estimated using adaptive algorithms by reducing a certain cost function. The techniques of gradient descent generally use convex and uni-modal cost functions to converge on the global minimum error surfaces. In this case, for better approximation of optimum error nonlinearity, a mixture of LMS and LMF families of algorithms can be used:

$$\nabla e^{*}(n) = -x^{*}(n) \tag{5.4}$$

where $\Delta_w \cdot F(e(n))$ is the first gradient of (5.), $\mu > 0$ is the step size, and $\alpha > 0$ is the design parameter.

Now, the stochastic gradient update is given by

$$w(n+1) = w(n) + \mu \cdot x(n) \frac{\partial f(e(n))}{\partial e(n)} \left[\frac{\alpha f(e(n))}{1 + \alpha f(e(n))} \right] \tag{5.5}$$

The algorithm is similar to an LMF error update, and it resembles LMS algorithm for high-error perturbations. This offers a lower MSE in steady state thanks to the fourth-order error statistics for small perturbations and the stability of least-squares algorithms for greater perturbations. The flow chart diagram for LMLS algorithm is as shown in Figure 3.7.

From the LMS weight recursion expression, $\mu(n)$ is the time variable step size parameter chosen so that a posteriori error

$$e^{*}(n) = d(n) - w^{T}(n+1)\, x(n) \tag{5.7}$$

is minimized in magnitude. The following expression results in minimizing $(e^{*}(n))^2$ with regard to $\mu(n)$ which forces $e^{*}(n)$ to zero:

$$\mu(n) = \frac{1}{2\, x^{T}(n)\, x(n)} \tag{5.8}$$

Thus, from (5.8) in (4.1), the resultant expression is obtained as

$$w(n+1) = w(n) + \frac{1}{2\, x^{T}(n)\, x(n)} e(n) x(n) \tag{5.9}$$

The adaptation constant μ' is dimensionless for the NLMLS filter, while for the LMS filter, it has reverse power dimensions. The NLMLS can be viewed as an LMS filter using time-variable step size parameter depending on relative logarithmic cost function. The normalized LMLS filter presents an issue of its own to resolve the problems of the gradient noise amplification and weight drift associated with LMS filter. In this case, numerical difficulties may occur, and then we must divide the squared norm $\|x_n\|^2$ by a small value.

Thus, the weight update equation of the NLMLS algorithm is given as

$$w(n+1) = w(n) + \mu(n).x(n)e(n) \left[\frac{\alpha (e(n))^2}{1 + \alpha (e(n))^2} \right] \tag{5.12}$$

where $\mu(n)=\dfrac{\mu'}{\varepsilon+\|x(n)\|^2}$ is a normalized step size with $0<\mu<2$.

The replacement of μ in the LMLS update vector weight equation with $\mu(n)$ from (5.10) contributes to the NLMLS, given by

$$w(n+1)=w(n)+\frac{\mu'}{\|x(n)\|2}\,x(n)e(n)\left[\frac{\alpha(e(n))^2}{1+\alpha(e(n))^2}\right] \tag{5.13}$$

The correction $w(n)$ in LMS is proportional to input vector $x(n)$, and a gradient noise amplification problem is encountered with LMS algorithm. However, this issue is diminished by normalization of LMS step size by $\|x(n)\|^2$ in NLMLS algorithm. Although the NLMLS algorithm bypasses this issue, we now face a similar issue that occurs while $\|x(n)\|$ is more lower. A framework based on logarithmic number system to implement adaptive filters with error nonlinearities in hardware is presented in Ref. [126].

Comparing (3.17) and (5.15), NLMLS update equation is a scaled LMLS algorithm variant. The change in the size of weight vector $w(n)$ is thus inversely proportional to the data vector $x(n)$. Data vector $x(n)$ with a larger norm usually leads to small changes in $w(n)$ than a smaller norm vector. This normalization results in smaller step size values than standard LMS and also exhibits weaker convergence compared to LMS algorithm because it uses a variable convergence factor to minimize instantaneous output error. Also, the step size parameter μ is independent of the signal input power. But to assess the normalization term $\|x(n)\|^2$, NLMLS algorithm requires less prior data than LMS at the same moment, and its MSE is thus higher than LMS. So as to further reduce the computational difficulty, we adopt block processing of normalized algorithms considered in this chapter. But at this time, we consider overlapping blocks. This method is also termed as maximum normalization, and the resulting algorithm is maximum NLMLS algorithm. The flow chart for the maximum form of NLMLS algorithm is as depicted in Figure 5.1.

Now, the weight update relation of maximum NLMLS (MNLMLS) algorithm for $x_{Li}\neq 0$ and $\varepsilon=0$ is written from (5.10) as

$$w(n+1)=w(n)+\frac{\mu'}{x_{Li}^2}x(n)\,e(n)\left[\frac{\alpha(e(n))^2}{1+\alpha(e(n))^2}\right] \tag{5.15}$$

Here, we have also considered the weight update relation for the sign regressor form of maximum SRNLMLS algorithm which is written from (5.10) as

$$w(n+1)=w(n)+\frac{\mu'}{x_{Li}^2}\mathrm{Sign}\{x(n)\}\,e(n)\left[\frac{\alpha(e(n))^2}{1+\alpha(e(n))^2}\right] \tag{5.16}$$

where $x_{Li}=\max\{|x_k|,k\in Z_i'\}$, $Z_i'=\{iL,\ iL+1,\ldots,iL+L-1\}$, $i\in Z$, and for $x_{Li}=0$ and $\varepsilon=0$, Equation (5.15) becomes $w(n+1)=w(n)$.

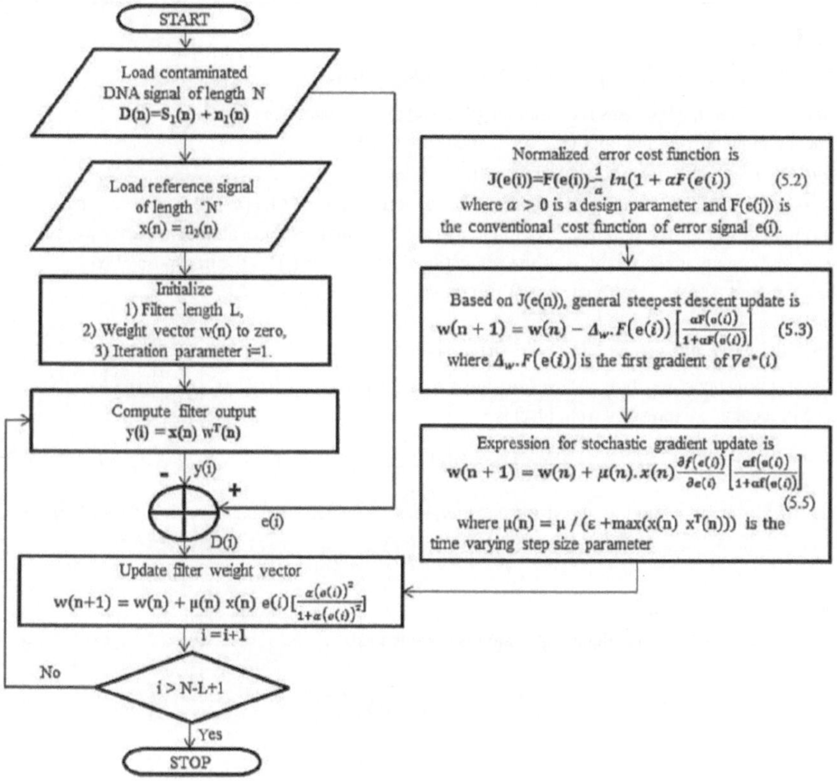

FIGURE 5.1 Flow chart diagram for maximum NLMLS algorithm.

5.4 ERROR-NORMALIZED LMLS (ENLMLS) ALGORITHM

Instantaneous data vector can be used to normalize the squared error vector norm in its place. The error vector length is the immediate number of iterations. The vector length of error is instantaneous. This algorithm is called the ENLMLS, as its step size with regard to error has been normalized. The step size of ENLMLS variable with regard to time remains inversely proportional to squared error vector norm rather than NLMLS information input vector [126]. This algorithm provides substantial changes to reduce noise distortion. ENLMLS algorithm has the benefit of selecting the step size regardless of input signal power and tap weights. Therefore, with its faster completion speed and steady-state error, the ENLMLS is faster than LMS (Table 5.3).

$w(n)$ is the tap-weight vector with the taps stored in a row vector

$$= \left[w(n)\ w(n-1)\ \ldots\ w(n-T+1)\right]^T$$

Here, an unknown vector ω_0 is presented with a linear model as

$$d(n) = \omega_0^T x(n) + n_t$$

TABLE 5.3
Mathematical Modeling of ENLMLS Algorithm

Parameters: T = number of taps (i.e. filter length), μ = step size parameter

Let the tap input be $x(n)$ and filter length T is moderate to large.

Initialization: Set $w(0) = 0$ as the initial condition.

Data: Given $x(n) = T$-by-1 tap input vector to filter n_2 at time $n = \left[x(n), x(n-1) \ldots x(n-T+1)\right]^T$, $w(n)$ is the tap-weight vector of adaptive filter, $d(n)$ is the desired response at time n, $e(n)$ is the error signal, ω_0 is an unknown vector, μ' is the adaptation constant, and $(.)^T$ is the transpose of (.).

To be computed: $w(n+1)$ = estimate of tap-weight vector at time $n+1$

Computation:

The FIR filter output is given by $y(n) = x^T(n)w(n) = w^T(n)x(n)$

The conventional cost function of the error signal $e(n)$ is $F\left[e(n)\right] = E\left[\left(e(n)\right)^2\right] = E\left[\left|e(n)\right|\right]$.

The LMS recursion expression from (4.1) is

$$w(n+1) = w(n) + 2\,\beta(n)\,x(n)\,e(n) \tag{5.18}$$

where $\beta(n)$ is the time-varying step size parameter, that is selected so that a posteriori error $e^*(n)$

$$e^*(n) = \left(1 - 2\,\beta(n)\,x^T(n)\,x(n)\,e(n)\right) \tag{5.19}$$

is minimized in magnitude.

From (5.20) in (5.18), the resultant expression is obtained asw

$$w(n+1) = w(n) + \frac{1}{2\,e^T(n)\,e(n)} e(n)x(n) \tag{5.21}$$

The bound on step size for mean square convergence of LMS is given as

$$0 < \beta' < \frac{2}{\boldsymbol{x}^T(n)\boldsymbol{x}(n)} \tag{5.22}$$

Setting $\beta(n)$ in the case of ENLMLS algorithm, it becomes

$$\beta(n) = \frac{\beta'}{e^T(n)e(n)} = \frac{\beta'}{\left\|e(n)\right\|^2} \tag{5.23}$$

A small positive constant ε is added to prevent denominator from becoming too low:

$$\beta(n) = \frac{\beta'}{\varepsilon + \left\|e(n)\right\|^2} \tag{5.24}$$

By replacing μ in the LMS weight vector update equation with $\beta(n)$ from (5.24), the weight update relation for ENLMLS is written as

$$\boldsymbol{w}(n+1) = \boldsymbol{w}(n) + \frac{\beta'}{\varepsilon + \left\|e(n)\right\|^2} \boldsymbol{x}(n) \left[\frac{\alpha\left(e(n)\right)^2}{1 + \alpha\left(e(n)\right)^2}\right] \tag{5.26}$$

Thus, the weight update relation for maximum ENLMLS for $e_{Li} \neq 0$ and $\varepsilon \neq 0$ becomes

$$\boldsymbol{w}(n+1) = \boldsymbol{w}(n) + \frac{\beta'}{\varepsilon + \max\left(\left\|e(n)\right\|\right)^2} \boldsymbol{x}(n) \left[\frac{\alpha\left(e(n)\right)^2}{1 + \alpha\left(e(n)\right)^2}\right] \tag{5.28}$$

where the input signal is denoted by $x(n)$ and the desired signal by $d(n)$, and n_t represents the noise.

An unknown system vector here is estimated using adaptive techniques by minimizing a certain cost function.

Reducing $(e^*(n))^2$ with regard to $\beta(n)$ gives the below expression:

$$\beta(n) = \frac{1}{2\, e^T(n)\, e(n)} \tag{5.20}$$

that forces $e^*(n)$ to zero.

ENLMLS filter may be seen as an LMS filter that uses step parameter varied in time. For ENLMLS technique, input tap vector $x(n)$ is small to prevent the drawback of LMS gradient noise amplification. Numerical difficulties may show up here, and also the squared norm $\|e(n)\|^2$ needs to be divided with a lower value. A small positive constant ε is used to prevent too lower divisor and larger parameter for step size.

At this instant, step parameter becomes

$$\beta(n) = \frac{\beta'}{\varepsilon + \|e(n)\|^2} \tag{5.24}$$

where $\beta(n)$ is a normalized size of step with $0 < \beta' < 2$. The adaptation constant β' is dimensionless for ENLMLS filter, whereas for the LMS filter, it uses reverse dimensions.

Replacing μ in NLMLS weight update expression with $\beta(n)$ from (5.23) leads to ENLMLS written as

$$w(n+1) = w(n) + \frac{\beta'}{\|e(n)\|^2} x(n) \left[\frac{\alpha(e(n))^2}{1 + \alpha(e(n))^2} \right] \tag{5.25}$$

Comparing (3.17) and (5.26), ENLMLS weight update expression is a scaled variant of NLMLS technique. Weight vector $w(n)$ change in terms of size is thus inversely proportional to error vector $e(n)$ norm. The static parameter β' is selected properly to accomplish a better balance among convergence speed and smaller final MSE $e(n)^2$. The value of the number of iterations n increases in (5.24), and thus, $\beta(n)$ is the diminishing function of n. The ENLMLS algorithm needs a small amount of computations, compared to other normalized algorithms.

For this, the value of error observed in initial iteration is stored and squared. This value is then squared and added to the prior stored value in the next iteration, and this continues. On the other side, additional calculations are necessary to calculate $\beta(n)$.

So as to further lower the computational difficulty, we have adopted block processing of error-normalized LMS and its sign regressor version with overlapping blocks in this chapter. The flow chart for the maximum form of ENLMLS is as depicted in Figure 5.2. Now the weight update relation of maximum ENLMLS (MENLMLS) algorithm for $e_{Li} \neq 0$ and $\varepsilon = 0$ is written from (5.23) as

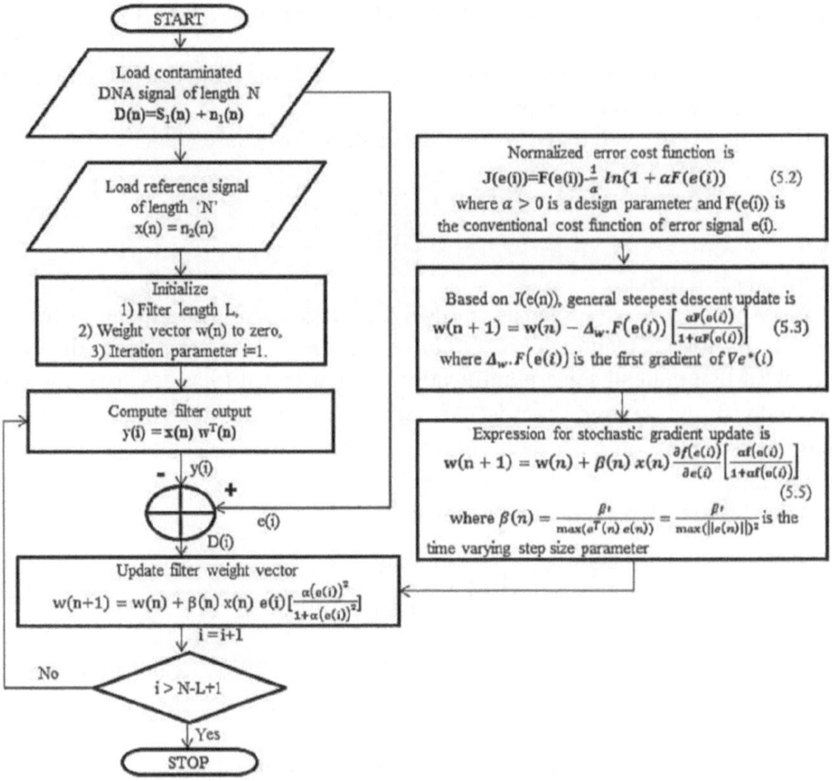

FIGURE 5.2 Flow chart diagram for maximum ENLMLS algorithm.

$$w(n+1) = w(n) + \frac{\beta'}{e_{Li}^2} x(n) \left[\frac{\alpha(e(n))^2}{1+\alpha(e(n))^2} \right] \tag{5.27}$$

where $e_{Li} = \max\{|e_k|, k \in Z_i'\}$, $Z_i' = \{iL,\ iL+1, \ldots, iL+L-1\}$, $i \in Z$, and for $e_{Li} = 0$ and $\varepsilon = 0$, Equation (5.27) becomes $w(n+1) = w(n)$.

5.5 NORMALIZED LLAD (NLLAD) ALGORITHM

In order to increase adaptive exon predictor (AEP) efficiency over LMS, normalized LLAD algorithms are taken into account, elegantly and steadily altering the cost function for optimizing errors. LLAD is robust and overcomes impulsive interference by outperforming SA [139]. Logarithmic LLAD variant based on normalization and relative logarithmic cost is referred to as NLLAD algorithm (Table 5.4).

$w(n)$ is the tap-weight vector with the taps stored in a row vector

$$= \left[w(n)\ w(n-1) \ldots w(n-T+1) \right]^T$$

TABLE 5.4
Mathematical Modeling of NLLAD Algorithm

Parameters: T = number of taps (i.e. filter length), μ = step size parameter

Let the tap input be $x(n)$ and filter length T is moderate to large.

Initialization: Set $w(0) = 0$ as the initial condition.

Data: Given $x(n) = T$-by-1 tap input random vector to filter n_2 at time $n = \left[x(n), x(n-1) \ldots x(n-T+1)\right]^T$ $w(n)$ is the tap-weight vector of adaptive filter, $d(n)$ is the desired response at time n, $e(n)$ is the error signal, ω_0 is an unknown vector, μ' is the adaptation constant, and $(.)^T$ is the transpose of (.).

To be computed: $w(n+1)$ = estimate of tap-weight vector at time $n+1$

Computation: The instantaneous estimate of gradient vector J is written as

$$\nabla J(n) = -\,2\,x(n)\,d^*(n) + 2\,d(n)\cdot x^T(n)\cdot w(n) \tag{5.29}$$

Substituting estimate in the steepest descent algorithm, new recursive relation to update tap-weight vector is obtained as

$$w(n+1) = w(n) + \mu\; x(n)\left[d^*(n) - x^T(n)\cdot w(n)\right] \tag{5.32}$$

SA utilizes $F(e(n)) = E\left[|e(n)|\right]$ as the cost function, which provides robustness against impulsive interference.

Here, the normalized error cost function introduced using logarithmic function is given by

$$J(e(n)) = F(e(n)) - \frac{1}{\alpha}\ln\left(1 + \alpha F(e(n))\right) \tag{5.33}$$

Thus, the cost function of $F(e(n))$ is written as

$$F\left[e(n)\right] = E\left[(e(n))^2\right] = E\left[|e(n)|\right] \tag{5.34}$$

where $\alpha > 0$ is a design parameter and $F(e(n))$ is a conventional cost function of error signal $e(n)$.

The stochastic gradient update expression is given by

$$w(n+1) = w(n) + \mu . x(n)\frac{\partial f(e(n))}{\partial e(n)}\left[\frac{\alpha f(e(n))}{1 + \alpha f(e(n))}\right] \tag{5.37}$$

Setting the time varying size of step parameter as

$$\mu(n) = \frac{\mu'}{\mathbf{x}^T(n)\mathbf{x}(n)} = \frac{\mu'}{\|\mathbf{x}(n)\|^2} \tag{5.40}$$

where $\mu(n)$ is a normalized step size with $0 < \mu' < 2$.

A small positive constant ε is added to prevent denominator from becoming too low:

$$\mu(n) = \frac{\mu'}{\varepsilon + \|\mathbf{x}(n)\|^2} \tag{5.42}$$

The alternate weight expression for NLLAD algorithm is written as

$$w(n+1) = w(n) + \frac{\mu'}{\varepsilon + \|\mathbf{x}(n)\|^2}\,x(n)\cdot\left[\frac{\alpha(e(n))}{1 + \alpha(|e(n)|)}\right] \tag{5.43}$$

Thus, the weight update relation for MNLLAD for $e_{Li} \neq 0$ and $\varepsilon \neq 0$ becomes

$$\mathbf{w}(n+1) = \mathbf{w}(n) + \frac{\mu'}{\varepsilon + \max\left(\|e(n)\|\right)^2}\,\mathbf{x}(n)\left[\frac{\alpha(e(n))}{1 + \alpha(|e(n)|)}\right] \tag{5.45}$$

Here, an unknown vector ω_0 is presented with a linear model as

$$d(n) = \omega_0^T x(n) + n_t$$

The expression for estimation error is given by

$$e(n) = d(n) - w^T(n).x(n) \tag{5.30}$$

where the term $w^T(n) \cdot x(n)$ is the inner product of *w*(*n*) and *x*(*n*).

The signum representation is written as

$$\text{Sign}\{x(n)\} = \left\{ \begin{array}{c} 1 : x(n) > 0 \\ 0 : x(n) = 0 \\ -1 : x(n) < 0 \end{array} \right\} \tag{5.31}$$

SA utilizes $F(e(n)) = E[|e(n)|]$ as the cost function, which provides robustness against impulsive interference.

For $|\alpha\ F(e(n))| \leq 1$, applying Maclaurin series using natural algorithm, Equation (5.34) gives

$$J(e(n)) = F(e(n)) - \frac{1}{\alpha}\left(\alpha F(e(n)) - \alpha^2\ F^2(e(n))\right) \tag{5.35}$$

For lesser values of $F(e(n))$, it is an infinite combination for conventional cost functions. For smaller values of error, the cost function produces the second power of $F(e(n))$, while for fairly big values of error, cost function *J*(*e*(*n*)) resemble $F(e(n))$ and is written as

$$\left[F\ (e\ (n)) - \frac{1}{\alpha}\ \text{ln}\left(1 + \alpha F(e(n))\right)\right] F(e(n)) \tag{5.36}$$

as *e*(*n*) tends to be an infinite value.

As the norm is the power of the least probable error aimed at convex cost function, SA delivers a slower convergence rate.

Thus, for $F[e(n)] = E\left[(e(n))^2\right]$, the weight update expression for LLAD algorithm becomes

$$w(n+1) = w(n) + \mu \cdot x(n)\left[\frac{\alpha(e(n))}{1 + \alpha(|e(n)|)}\right] \tag{5.38}$$

Thus, the weight update equation of the NLLAD algorithm is given as

$$w(n+1) = w(n) + \mu(n) \cdot x(n).\left[\frac{\alpha(e(n))}{1 + \alpha(|e(n)|)}\right] \tag{5.39}$$

To upsurge the convergence and stability of AEP, NLLAD based on relative logarithmic cost is presented. NLLAD algorithm overcomes LMS setbacks and also enhances tracking ability along with convergence speed. The flow chart for NLLAD algorithm is as shown in Figure 5.3.

Thus, the weight update equation of NLLAD algorithm is given as

$$w(n+1) = w(n) + \frac{\mu'}{\left(\|x(n)\|\right)^2}\, x(n).\left[\frac{\alpha(e(n))}{1+\alpha\left(|e(n)|\right)}\right] \tag{5.41}$$

where $\mu(n) = \frac{\mu'}{\|x(n)\|^2}$ is a normalized step size with $0 < \mu' < 2$.

So as to overcome gradient noise amplification problem and weight drift problem linked with standard LMS, NLLAD-based AEP is used. The correction to $w(n)$ in LMS is proportional to $x(n)$, and also LMS has an issue with amplification of gradient noise. This noise amplification problem decreases with use of normalized step size of NLLAD by $\|x(n)\|^2$. While the NLLAD algorithm bypasses this issue, we now face a comparable issue that occurs when $\|x(n)\|$ remains too low. Comparing (3.17) and (5.40), NLLAD update equation is a scaled LLAD variant. Thus, the size of the

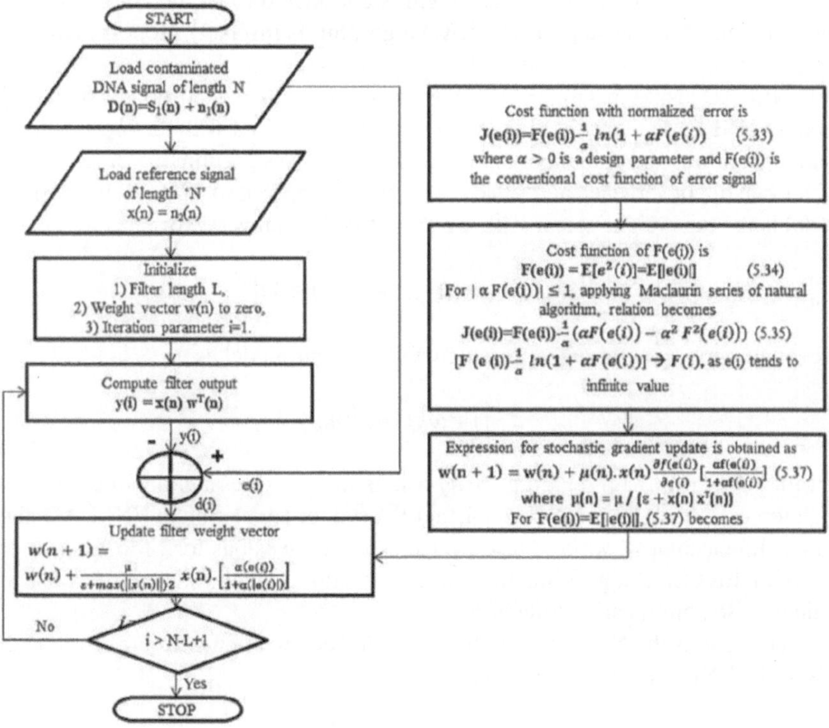

FIGURE 5.3 Flow chart diagram for maximum NLLAD algorithm.

change to $w(n)$ is inversely relational to the data vector norm of $x(n)$. Data vector with a larger norm is usually a minor change to $w(n)$ than a smaller norm vector.

This normalization results in smaller step size values and also presents quicker convergence than LMS because it uses a variable convergence factor that minimizes instantaneous output errors. To evaluate normalized term $\|\boldsymbol{x}(n)\|^2$, NLLAD needs less information than LMS at the same moment. To reduce the computational complexity, we adopt block processing of NLLAD algorithm with use of overlapping blocks. The flow chart for maximum NLLAD (MNLLAD) algorithm is as depicted in Figure 5.3.

Now, the weight update relation of MNLLAD algorithm for $e_{Li} \neq 0$ and $\varepsilon = 0$ is written from (5.40) as

$$\boldsymbol{w}(n+1) = \boldsymbol{w}(n) + \frac{\mu'}{e_{Li}^2}\boldsymbol{x}(n)\left[\frac{\alpha(e(n))}{1+\alpha(|e(n)|)}\right] \tag{5.44}$$

where $e_{Li} = \max\{|e_k|, k \in Z_i'\}$, $Z_i' = \{iL,\ iL+1,\ldots,iL+L-1\}$, $i \in Z$, and for $e_{Li} = 0$ and $\varepsilon = 0$, Equation (5.42) becomes $\boldsymbol{w}(n+1) = \boldsymbol{w}(n)$.

5.6 ERROR-NORMALIZED LLAD (ENLLAD) ALGORITHM

An error vector squared norm using normalization is used here rather than an instantaneous vector for data. Error vector length is considered to be the amount of iterations. The time-varying step in ENLLAD algorithm is inversely proportional to the squared norm of error vector and not the data input vector as shown in NLLAD algorithm. NLLAD also offers major enhancements to minimize distortion of signal. The benefit of ENLLAD algorithm is that size of step is selected without dependency on signal input power and amount of tap weights. ENLLAD algorithm therefore provides a better rate of convergence and steady-state error compared to LMS (Table 5.5).

$w(n)$ is the tap-weight vector with the taps stored in a row vector

$$= [w(n)\ w(n-1)\ \ldots\ w(n-T+1)]^T$$

Here, an unknown vector ω_0 is presented with a linear model as

$$d(n) = \omega_0^T x(n) + n_t$$

Minimizing $(e^*(n))^2$ with regard to $\beta(n)$ results in the expression obtained in (5.20) which forces $e^*(n)$ to be zero. The constant β' used for adaptation in ENLLAD algorithm is dimensionless, while it has reverse power dimensions for LMS filter. In the event of an ENLLAD algorithm, tap input vector $x(n)$ is small to overcome the LMS gradient noise amplification drawback.

Replacing μ in the NLLAD weight vector update equation with $\beta(n)$ from (5.50) leads to the ENLLAD, which is given as

$$w(n+1) = w(n) + \frac{\beta'}{\|e(n)\|^2}\, x(n)\left[\frac{\alpha(e(n))}{1+\alpha(|e(n)|)}\right] \tag{5.52}$$

TABLE 5.5
Mathematical Modeling of ENLLAD Algorithm

Parameters: T = number of taps (i.e. filter length), μ = step size parameter

Let the tap input be $x(n)$ and filter length T is moderate to large.

Initialization: Set $w(0) = 0$ as the initial condition.

Data: Given $x(n) = T$-by-1 tap input vector to filter n_2 at time $n = \left[x(n), x(n-1) \ldots x(n-T+1)\right]^T$

$w(n)$ is the tap weight vector of adaptive filter, $d(n)$ is the desired response at time n, $e(n)$ is the error signal, ω_0 is an unknown vector, μ' is the adaptation constant, and $(.)^T$ is the transpose of (.).

To be computed: $w(n+1)$ = estimate of tap-weight vector at time $n+1$

Computation: The FIR filter output is given by $y(n) = x^T(n)w(n) = w^T(n)x(n)$.

The conventional cost function of the error signal $e(n)$ is $F\left[e(n)\right] = E\left[\left(e(n)\right)^2\right] = E\left[\left|e(n)\right|\right]$.

The LMS recursion expression from (5.18) is written as

$$w(n+1) = w(n) + 2\ \beta(n)\ x(n)\ e(n) \tag{5.46}$$

where $\beta(n)$ is the time-varying step size parameter, that is selected so that a posteriori error $e^*(n)$ becomes

$$e^*(n) = \left(1 - 2\ \beta(n) x^T(n) x(n) e(n)\right) \tag{5.47}$$

is minimized in magnitude.

From (5.20) in (5.46), the resultant expression is obtained as

$$w(n+1) = w(n) + \frac{1}{2\ e^T(n)\ e(n)} e(n) x(n) \tag{5.48}$$

The limit of the step size for LMS mean square convergence is specified as

$$0 < \beta' < \frac{2}{\mathbf{x}^T(n)\mathbf{x}(n)} \tag{5.49}$$

Also, setting $\beta(n)$ in the case of ENLLAD algorithm, it becomes

$$\beta(n) = \frac{\beta'}{e^T(n)\ e(n)} = \frac{\beta'}{\left\|e(n)\right\|^2} \tag{5.50}$$

A small positive constant parameter ε is added to prevent denominator from becoming too low:

$$\beta(n) = \frac{\beta'}{\varepsilon + \left\|e(n)\right\|^2} \tag{5.51}$$

By replacing μ in the NLLAD weight vector update equation with $\beta(n)$ from (5.51), weight update relation for ENLLAD is written as

$$\mathbf{w}(n+1) = \mathbf{w}(n) + \frac{\beta'}{\varepsilon + \left\|e(n)\right\|^2} \mathbf{x}(n) \left[\frac{\alpha(e(n))}{1 + \alpha\left(\left|e(n)\right|\right)}\right] \tag{5.53}$$

Thus, the weight update relation for maximum ENLMLS for $e_{Li} \neq 0$ and $\varepsilon \neq 0$ becomes

$$\mathbf{w}(n+1) = \mathbf{w}(n) + \frac{\beta'}{\varepsilon + \max\left(\left\|e(n)\right\|\right)^2} \mathbf{x}(n) \left[\frac{\alpha(e(n))}{1 + \alpha\left(\left|e(n)\right|\right)}\right] \tag{5.55}$$

where $\beta(n)$ is a normalized step size with $0 < \beta' < 2$.

Squared norm $\|e(n)\|^2$ must be divided with a low value to prevent numerical issues by adding a small positive constant ε. Comparing (3.17) and (5.53), weight update relation of ENLLAD is a scaled variant of NLLAD. Consequently, the size of the change to $w(n)$ is inversely relational to vector $e(n)$. The specified parameter β needs to be properly selected to achieve the best trade-off among the speed of convergence and low final MSE $e(n)^2$ in (5.24), as it increases with the number of iterations 'n' and thus $\beta(n)$ is a decreasing function of n. ENLLAD offers lesser computing difficulty compared to other normalized algorithms. The error value created in the initial iteration will be squared and stored to calculate (5.51) with low computing difficulty, then squared and added to its previous values, and this continues. Additional computations to measure $\beta(n)$, on other side, are needed. For further minimizing computing difficulty, we have adopted block processing of error-normalized LLAD and its maximum version with overlapping blocks at this time in this section. The flow chart for the maximum form of ENLLAD algorithm is as depicted in Figure 5.4.

Now, the weight update relation of MENLLAD algorithm for $e_{Li} \neq 0$ and $\varepsilon = 0$ is written from (5.50) as

$$w(n+1) = w(n) + \frac{\beta'}{e_{Li}^2} x(n) \left[\frac{\alpha(e(n))}{1+\alpha(\|e(n)\|)} \right] \tag{5.54}$$

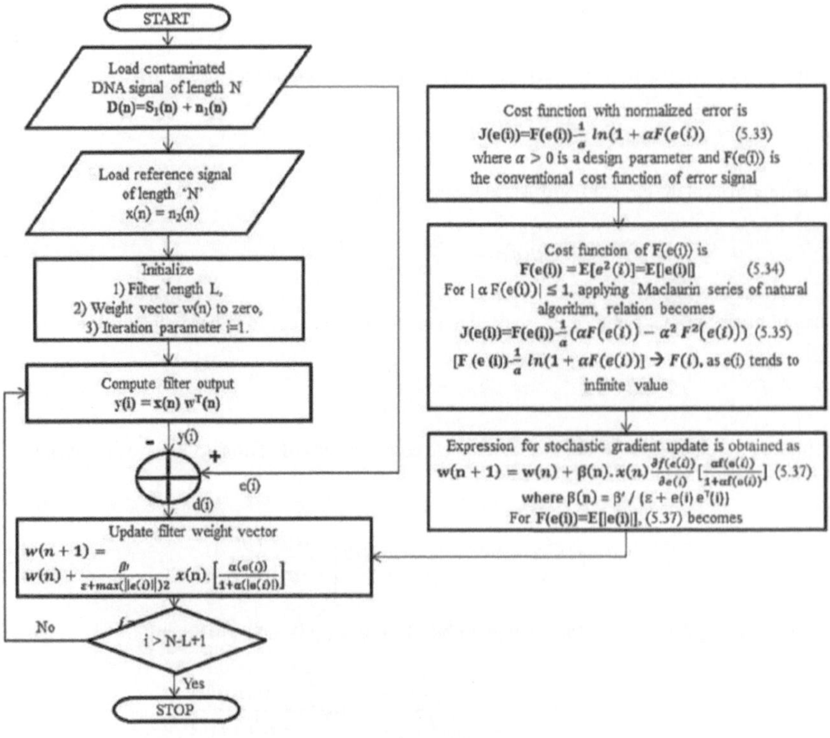

FIGURE 5.4 Flow chart diagram for maximum ENLLAD algorithm.

where $e_{Li} = \max\{|e_k|, k \in Z_i'\}, Z_i' = \{iL,\ iL+1, \ldots, iL+L-1\}, i \in Z$, and for $e_{Li} = 0$ and $\varepsilon = 0$, Equation (5.51) becomes $\boldsymbol{w}(n+1) = \boldsymbol{w}(n)$.

5.7 EXTENSION TO SIGN-BASED REALIZATIONS OF LOGARITHMIC NORMALIZED ALGORITHMS

In bioinformatics applications, low computational complexity of an adaptive filter is extremely desirable. This reduction is usually achieved by either clipping input or estimating error. Algorithms based on error and data clipping are provided in Section 3.9, and these are SRA, SA, and SSA. These three simpler algorithms combined with logarithmic normalized adaptive techniques provide quicker convergence and a decreased computational difficulty.

5.7.1 Extension to Sign-Based Realizations of NLMLS-Based Variants

NLMLS algorithm is a basic version of higher-order adaptive filter [127]. Combining the NLMLS in (5.14) with SRA, SA, and SSA results in its signed variants NSRLMLS, NSLMLS, and NSSLMLS, respectively. Their weight expressions are given as follows:

$$w(n+1) = w(n) + \frac{\mu'}{\varepsilon + \|x(n)\|^2} \operatorname{Sign}\{x(n)\}\, e(n)\left[\frac{\alpha(e(n))^2}{1+\alpha(e(n))^2}\right] \quad (5.56)$$

$$w(n+1) = w(n) + \frac{\mu'}{\varepsilon + \|x(n)\|^2}\, x(n) \operatorname{Sign}\left\{e(n)\left[\frac{\alpha(e(n))^2}{1+\alpha(e(n))^2}\right]\right\} \quad (5.57)$$

$$w(n+1) = w(n) + \frac{\mu'}{\varepsilon + \|x(n)\|^2} \operatorname{Sign}\{x(n)\} \operatorname{Sign}\left\{e(n)\left[\frac{\alpha(e(n))^2}{1+\alpha(e(n))^2}\right]\right\} \quad (5.58)$$

From (5.57) and (5.58), it is noted that the weight update relations for NSLMLS and NSSLMLS are the same as SLMS and SSLMS. Among these, the NSRLMLS algorithm in (4.56) provides better convergence performance and good tracking ability due to its low computational difficulty. To further minimize computing difficulty, we have adopted block processing of sign-based normalized LMLS techniques by considering overlapping blocks. This method is also termed as maximum normalization.

The resulting algorithms include MNLMLS, MNSRLMLS, MNSLMLS, and MNSSLMLS.

The weights update relations of MNSRLMS, MNSLMLS, and MNSSLMLS algorithms for $x_{Li} \neq 0$ and $\varepsilon \neq 0$ are written as

$$w(n+1) = w(n) + \frac{\mu'}{\varepsilon + \max\left(\|x(n)\|\right)^2} \operatorname{Sign}\{x(n)\}\, e(n)\left[\frac{\alpha(e(n))^2}{1+\alpha(e(n))^2}\right] \quad (5.56)$$

$$w(n+1)=w(n)+\frac{\mu'}{\varepsilon+\max\left(\left\|x(n)\right\|\right)^2}x(n)\ \text{Sign}\left\{e(n)\left[\frac{\alpha(e(n))^2}{1+\alpha(e(n))^2}\right]\right\} \tag{5.57}$$

$$w(n+1)=w(n)+\frac{\mu'}{\varepsilon+\max\left(\left\|x(n)\right\|\right)^2}\text{Sign}\{x(n)\}\ \text{Sign}\left\{e(n)\left[\frac{\alpha(e(n))^2}{1+\alpha(e(n))^2}\right]\right\} \tag{5.58}$$

Among these normalized and its maximum variants, the MNSRLMLS algorithm in (5.56) performs better than LMS and all its other variants due to its low computational complexity and good exon tracking ability.

5.7.2 Extension to Sign-Based Realizations of ENLMLS-Based Variants

ENLMLS algorithm is a basic version of higher-order adaptive filter. Combining the ENLMLS in (5.26) with SRA, SA, and SSA results in its signed variants ENSRLMLS, ENSLMLS, and ENSSLMLS, respectively. Their weight expressions are given as follows:

$$w(n+1)=w(n)+\frac{\beta'}{\varepsilon+\left\|e(n)\right\|^2}\ \text{Sign}\{x(n)\}\ e(n)\left[\frac{\alpha(e(n))^2}{1+\alpha(e(n))^2}\right] \tag{5.59}$$

$$w(n+1)=w(n)+\frac{\beta'}{\varepsilon+\left\|e(n)\right\|^2}\ x(n)\ \text{Sign}\left\{e(n)\left[\frac{\alpha(e(n))^2}{1+\alpha(e(n))^2}\right]\right\} \tag{5.60}$$

$$w(n+1)=w(n)+\frac{\beta'}{\varepsilon+\left\|e(n)\right\|^2}\ \text{Sign}\{x(n)\}\ \text{Sign}\left\{e(n)\left[\frac{\alpha(e(n))^2}{1+\alpha(e(n))^2}\right]\right\} \tag{5.61}$$

From (5.60) and (5.61), it is noted that the weight update relations for ENSLMLS and ENSSLMLS are the same as SLMS and SSLMS. Among these, the ENSRLMLS algorithm in (5.59) performs better due to its low computational difficulty, better convergence performance, and good tracking ability. To further minimize computing difficulty, we have adopted block-based method for sign-based ENLMLS techniques by considering overlapping blocks.

The additional computations required for computing the variable step size for all the ENLMLS-based algorithms discussed in the previous sections can be further reduced by using block approach in which data input is divided into blocks and magnitude is maximum in each block remains used for computing variable step size parameter. The resulting algorithms include MENLMLS, MENSRLMLS, MENSLMLS, and MENSSLMLS.

Thus, the weight update relations of MENSRLMS, MENSLMLS, and MENSSLMLS algorithms for $x_{Li} \neq 0$ and $\varepsilon \neq 0$ are written as

$$w(n+1) = w(n) + \frac{\beta'}{\varepsilon + \max\left(\left\|e(n)\right\|\right)^2} \operatorname{Sign}\{x(n)\}\, e(n)\left[\frac{\alpha(e(n))^2}{1+\alpha(e(n))^2}\right] \tag{5.62}$$

$$w(n+1) = w(n) + \frac{\beta'}{\varepsilon + \max\left(\left\|e(n)\right\|\right)^2} x(n) \operatorname{Sign}\left\{e(n)\left[\frac{\alpha(e(n))^2}{1+\alpha(e(n))^2}\right]\right\} \tag{5.63}$$

$$w(n+1) = w(n) + \frac{\beta'}{\varepsilon + \max\left(\left\|e(n)\right\|\right)^2} \operatorname{Sign}\{x(n)\} \operatorname{Sign}\left\{e(n)\left[\frac{\alpha(e(n))^2}{1+\alpha(e(n))^2}\right]\right\} \tag{5.64}$$

Among these normalized and its maximum variants, the MENSRLMLS algorithm in (5.62) performs better than LMS and all its other variants due to its low computational complexity and good exon tracking ability.

5.7.3 Extension to Sign-Based Realizations of NLLAD-Based Variants

NLLAD algorithm is a basic version of higher-order adaptive filter. Combining the NLLAD in (5.43) with SRA, SA, and SSA results in its signed variants NSRLLAD, NSLLAD, and NSSLLAD, respectively. Their weight expressions are given as follows:

$$w(n+1) = w(n) + \frac{\mu'}{\varepsilon + \left\|\boldsymbol{x}(n)\right\|^2} \operatorname{Sign}\{x(n)\}.\left[\frac{\alpha(e(n))}{1+\alpha\left(\left|e(n)\right|\right)}\right] \tag{5.65}$$

$$w(n+1) = w(n) + \frac{\mu'}{\varepsilon + \left\|\boldsymbol{x}(n)\right\|^2} x(n)\cdot \operatorname{Sign}\left\{\left[\frac{\alpha(e(n))}{1+\alpha\left(\left|e(n)\right|\right)}\right]\right\} \tag{5.66}$$

$$w(n+1) = w(n) + \frac{\mu'}{\varepsilon + \left\|\boldsymbol{x}(n)\right\|^2} \operatorname{Sign}\{x(n)\}\cdot \operatorname{Sign}\left\{\left[\frac{\alpha(e(n))}{1+\alpha\left(\left|e(n)\right|\right)}\right]\right\} \tag{5.67}$$

From (5.66) and (5.67), it is noted that the weight update relations for NSLLAD and NSSLLAD are the same as SLMS and SSLMS. Among these, the NSRLLAD algorithm in (5.65) performs better because of its less complexity to perform computations, better convergence performance, and good tracking ability [128]. To further lessen computing difficulty, block approach is used for sign-based normalized LLAD techniques by considering overlapping blocks.

The additional computations required for computing the variable step size for all the NLLAD-based algorithms discussed in the previous sections is further decreased thru this block approach, which is also termed as maximum normalization. The resulting algorithms include MNLLAD, MNSRLLAD, MNSLLAD, and MNSSLLAD.

Thus, the weight update relations of MNSRLLAD, MNSLLAD, and MNSSLLAD algorithms for $x_{Li} \neq 0$ and $\varepsilon \neq 0$ are written as

$$w(n+1) = w(n) + \frac{\mu'}{\varepsilon + \max\left(\left\|\boldsymbol{x}(n)\right\|\right)^2} \text{ Sign}\{x(n)\}\left[\frac{\alpha(e(n))}{1+\alpha\left(\left|e(n)\right|\right)}\right] \tag{5.68}$$

$$w(n+1) = w(n) + \frac{\mu'}{\varepsilon + \max\left(\left\|\boldsymbol{x}(n)\right\|\right)^2} x(n) \text{ Sign}\left\{\left[\frac{\alpha(e(n))}{1+\alpha\left(\left|e(n)\right|\right)}\right]\right\} \tag{5.69}$$

$$w(n+1) = w(n) + \frac{\mu'}{\varepsilon + \max\left(\left\|\boldsymbol{x}(n)\right\|\right)^2} \text{Sign}\{x(n)\}\text{Sign}\left\{\left[\frac{\alpha(e(n))}{1+\alpha\left(\left|e(n)\right|\right)}\right]\right\} \tag{5.70}$$

Among these normalized and maximum variants, the MNSRLLAD algorithm in (5.68) performs better than LMS and all its other variants due to its low computational complexity and good exon tracking ability.

5.7.4 Extension to Sign-Based Realizations of ENLLAD-Based Variants

ENLLAD algorithm is a basic version of higher-order adaptive filter. Combining the ENLLAD in (5.53) with SRA, SA, and SSA results in its signed variants ENSRLLAD, ENSLLAD, and ENSSLLAD, respectively. Their weight expressions are given below:

$$w(n+1) = w(n) + \frac{\beta'}{\varepsilon + \left\|e(n)\right\|^2} \text{ Sign}\{x(n)\} \cdot \left[\frac{\alpha(e(n))}{1+\alpha\left(\left|e(n)\right|\right)}\right] \tag{5.71}$$

$$w(n+1) = w(n) + \frac{\beta'}{\varepsilon + \left\|e(n)\right\|^2} x(n) \cdot \text{Sign}\left\{\left[\frac{\alpha(e(n))}{1+\alpha\left(\left|e(n)\right|\right)}\right]\right\} \tag{5.72}$$

$$w(n+1) = w(n) + \frac{\beta'}{\varepsilon + \left\|e(n)\right\|^2} \text{ Sign}\{x(n)\} \cdot \text{Sign}\left\{\left[\frac{\alpha(e(n))}{1+\alpha\left(\left|e(n)\right|\right)}\right]\right\} \tag{5.73}$$

From (5.72) and (5.73), it is noted that the weight update relations for ENSLLAD and ENSSLLAD are the same as SLMS and SSLMS. Among these, the ENSRLLAD algorithm in (5.71) performs better because of its low computing difficulty, better rate of convergence, and good tracking ability. To further diminish difficulty to perform computations, block method is used for sign-based error-normalized LLAD techniques by considering overlapping blocks. This normalization method is also known as maximum normalization. The resulting algorithms include MENLLAD, MENSRLLAD, MENSLLAD, and MNSSLLAD.

Thus, the weight update relations of MENSRLLAD, MENSLLAD, and MENSSLLAD algorithms for $x_{Li} \neq 0$ and $\varepsilon \neq 0$ are written as

$$w(n+1) = w(n) + \frac{\beta'}{\varepsilon + \max\left(\|e(n)\|\right)^2} \operatorname{Sign}\{x(n)\}\left[\frac{\alpha(e(n))}{1+\alpha\left(|e(n)|\right)}\right] \tag{5.74}$$

$$w(n+1) = w(n) + \frac{\beta'}{\varepsilon + \max\left(\|e(n)\|\right)^2} x(n) \cdot \operatorname{Sign}\left\{\left[\frac{\alpha(e(n))}{1+\alpha\left(|e(n)|\right)}\right]\right\} \tag{5.75}$$

$$w(n+1) = w(n) + \frac{\beta'}{\varepsilon + \max\left(\|x(n)\|\right)^2} \operatorname{Sign}\{x(n)\} \operatorname{Sign}\left\{\left[\frac{\alpha(e(n))}{1+\alpha\left(|e(n)|\right)}\right]\right\} \tag{5.76}$$

Among these normalized and maximum variants, the MENSRLLAD algorithm in (5.74) performs better than LMS and all its other variants due to its low computational complexity and good exon tracking ability.

5.8 COMPUTATIONAL COMPLEXITY ISSUES

In this section, we focus on presenting a comparison of computational complexities among various normalization-based logarithmic adaptive algorithms. All logarithmic normalized adaptive algorithms and their sign-based realizations with computational complexity figures are summarized in Table 5.6. The signed realizations offer a significant reduction in the amount of computations needed for NLMLS algorithm. Moreover, since its algorithm based on the sign regressor is largely free of multiplication operations, it provides an elegant means of analyzing genomic sequences. NLMLS algorithm is more complicated among normalized algorithms as it needs 2T+7 multiplication operations for computing weight update relation in (5.14), where L is the filter length. Similarly, ENLMLS also requires 2T+7 multiplication and 2T+2 addition operations. In the case of SRA, just single multiply computation is needed for calculating $\mu \cdot e(n)$.

There are no multiplications required for other two NLMLS techniques when selecting μ as power of two. Multiplication becomes a shift computation that is less complex in practical accomplishments. Here, within SSA, signum function is used for both data and vector, and μ is added to the weight vector using addition with sign check (ASC) operation. From Table 4.6, it is clear that the number of computations required for NSRLMLS and ENSRLMLS algorithms is independent of filter length. Both these algorithms required a less number of multiplications and one division, because of the sign term applied to data vector, i.e., all the elements of the data vector become −1 or 0 or 1, so that multiplications are greatly reduced.

The LLAD algorithm provides significant improvements in terms of low computations and convergence performance compared to LMS. It is evident that NLLAD algorithm needs only 2T+4 multiplication operations with lesser additions than NLMLS. However, both the NLMLS and NLLAD algorithms offer low

TABLE 5.6
Computational Complexity Comparison of Normalized and Maximum Normalized Logarithmic Adaptive Algorithms

S.		Without Maximum Normalization			With Maximum Normalization		
No.	Algorithm	Multiplications	Additions	Divisions	Multiplications	Additions	Divisions
1	NLMLS	2T+7	2T+2	1	T+7	T+2	1
2	NSRLMLS	2T	2T+2	1	T	T+2	1
3	NSLMLS	2T+5	2T+2	1	T+5	T+2	1
4	NSSLMLS	T+2	T+2	1	2	2	1
5	ENLMLS	2T+7	2T+2	1	T+7	T+2	1
6	ENSRLMLS	2T	2T+2	1	T	T+2	1
7	ENSLMLS	2T+5	2T+2	1	T+5	T+2	1
8	ENSSLMLS	T+2	T+2	1	2	2	1
9	NLLAD	2T+4	2T+2	1	T+4	T+2	1
10	NSRLLAD	T+4	2T+2	1	4	T+2	1
11	NSLLAD	2T+3	2T+2	1	T+3	T+2	1
12	NSSLLAD	T+3	T+2	1	3	2	1
13	ENLLAD	2T+4	2T+2	1	T+4	T+2	1
14	ENSRLLAD	T+4	2T+2	1	4	T+2	1
15	ENSLLAD	2T+3	2T+2	1	T+3	T+2	1
16	ENSSLLAD	T+3	T+2	1	3	2	1

computational difficulty compared to all other LMS-based variants. Among these two algorithms, NLLAD algorithm offers low complexity in performing the computations and is more suitable in real bioinformatics-based applications.

NSRLMLS algorithm is clearly capable of accurately locating exons in the PSD plot and also offers less computing difficulty with 2T multiplications. The signed variants of NLLAD are regarded to further enhance convergence efficiency. To further reduce the computational complexity, we have also considered signed variants of all normalization- and maximum normalization-based logarithmic adaptive algorithms as shown in Table 5.6. From these computations, it is vibrant that NSRLLAD-based technique needs only T+4 multiplications compared to all other signed LMS-based techniques.

Hence, AEP based on this algorithm may therefore be more appropriate in clinical remote health care and bioinformatics systems. Due to clipping and logarithmic cost function, NSRLLAD needs only four multiplications than LMS and all other normalized variants.

It is because of the presence of sign function in its weight update expression. However, the signal quality will be decreased due to clipping of data and error vectors. Therefore, NSSLLAD cannot be a better candidate for applications of genomic signal processing. Based on these factors and low calculations, it is true that NSRLLAD is a better candidate for precisely locating the exon segments in DNA in genomics applications. To significantly reduce the computational complexity, we have also extended to block processing of all normalized logarithmic algorithms and their signed versions.

Their resulting maximum normalization-based adaptive algorithms include MNLMLS, MNSRLMLS, MNSLMLS, MNSSLMLS, MENLMLS, MENSRLMLS, MENSLMLS, MENSSLMLS, MNLLAD, MNSRLLAD, MNSLLAD, MNSSLLAD, MENLLAD, MENSRLLAD, MENSLLAD, and MENSSLLAD. Among all these maximum normalized based and their signed adaptive algorithms, all the algorithms perform better than LMS algorithms, whereas MNSRLLAD algorithm outperforms all the normalized variants in terms of low computational complexity and better tracking ability.

5.9 CONVERGENCE ANALYSIS

Convergence curves are drawn between MSE and the number of iterations. MSE shall be calculated using the expression, i.e. $J = E\left\{\left|e(n)\right|^2\right\}$, in (3.8). The simulations for all the convergence curves are done using MATLAB. In this code, MSE is calculated for each sample for 4,000 iterations, and the average value is taken for the characterization. These curves are obtained using various normalized based AEPs exon predictions with a finite length adaptive filter, a random variance of 0.01, and a step size of 0.01. The convergence curves for distinct normalized adaptive algorithms discussed in the previous sections are shown in Figure 4.5. From the convergence plots, it is obvious that NLLAD outperforms among all the normalized based adaptive algorithms. Thus, the performances of all NLMLS, ENLMLS, NLLAD, and ENLLAD algorithms are comparable to each other and superior to LMS.

Due to the normalization factor, the performance of sign regressor version of all maximum normalized logarithmic adaptive algorithms is slightly inferior to the corresponding non-signed version of the algorithm; their corresponding convergence curves are depicted in Figures 5.6–5.9. But, due to its significantly lower computational complexity, MNSRLLAD algorithm offers better exon tracking ability and is suitable for applications in real time. Comparing the convergence performance of all normalized logarithmic algorithms with LMS variants in Figure 3.10, it is obvious that all normalized logarithmic algorithms offer quicker convergence.

The normalized algorithms offer important improvements in reducing signal distortion. NLMLS and ENLMLS algorithms obviously converge faster than LMS and NLMS algorithms. The parameter α is used to actually control convergence. If there is a large error, α tends towards unity and there will be very swift convergence. Likewise, if the error is low, α is small and slower convergence makes step size small. This actually happens when the adaptive filter is stable. This actually occurs when the adaptive filter is reaching the steady state. To further improve the convergence performance, NLLAD and ENLLAD adaptive algorithms are considered. Among all normalized based algorithms, it is evident from Figure 5.5 that NLLAD offers a better convergence performance compared to LMS and its all other normalized based algorithms. The convergence curves result from plotting the MSE over several samples.

To significantly reduce the computational complexity and for better convergence performance, we have further extended all these normalized algorithms to maximum normalized based adaptive algorithms. The resulting algorithms include MNLMLS,

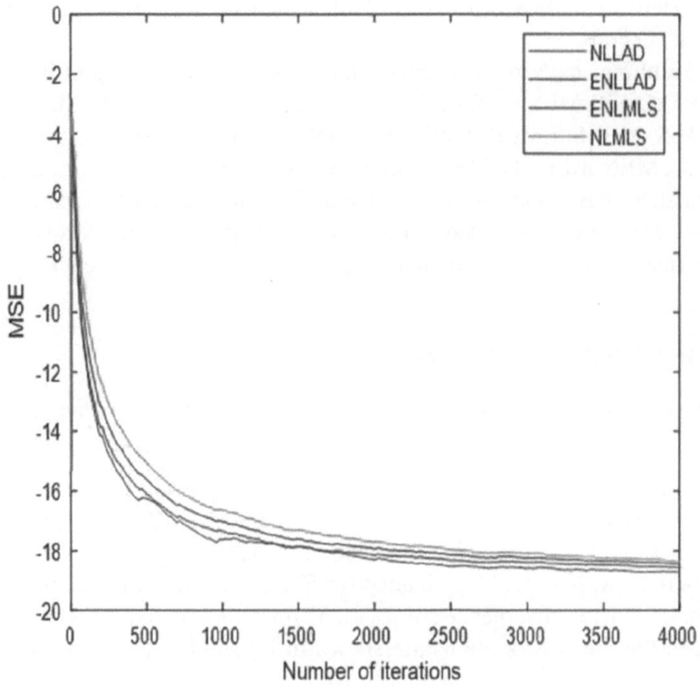

FIGURE 5.5 Convergence curves of normalized logarithmic adaptive algorithms.

MENLMLS, MNLLAD, and MENLLAD algorithms and their sign-based variants discussed in the previous sections. Figures 5.6–5.9 show the convergence curves for all maximum normalization-based logarithmic adaptive algorithms and their signed variants.

From all these curves, it is clear that the convergence performance of all MNLMLS, MENLMLS, MNLLAD, and MENLLAD algorithms converges faster than LMS by adopting maximum normalization as it uses a convergence variable factor to minimize instantaneous output error. Thus, MNLLAD algorithm is thus better than any other maximum normalized adaptive algorithms among the algorithms considered for implementing AEP for better convergence and exon locating capability.

5.10 RESULTS AND DISCUSSION FOR LOGARITHMIC NORMALIZED VARIANTS

In this section, performances of various maximum normalized logarithmic-based AEPs presented in this chapter are analyzed. AEP arrangement is as shown in Figure 3.2. Various maximum normalized based algorithms including MNLMLS, MENLMLS, MNLLAD, and MENLLAD and also their signed variants are used to develop multiple AEPs. LMS-based AEP is also implemented for comparison. For the purpose of assessment, the National Center for Biotechnology Information

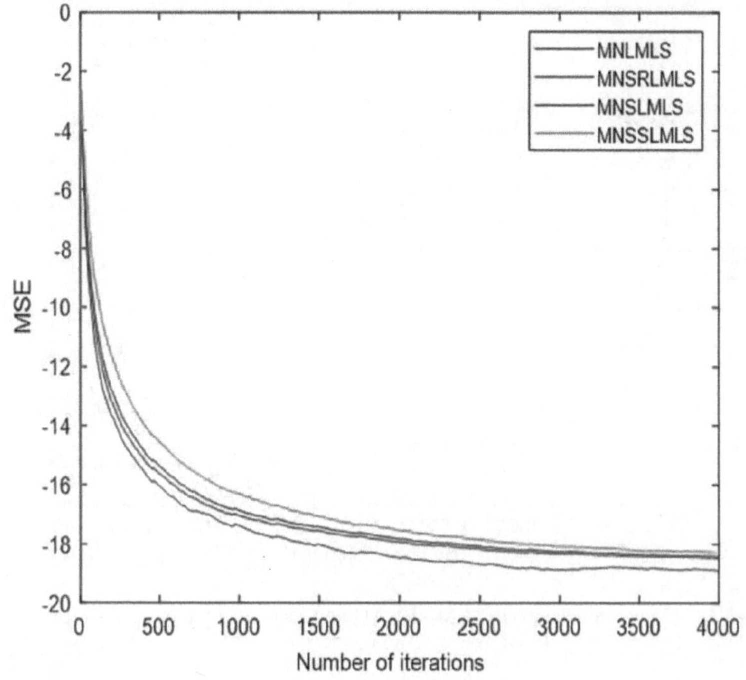

FIGURE 5.6 Convergence curves of MNLMLS and its signed adaptive algorithms.

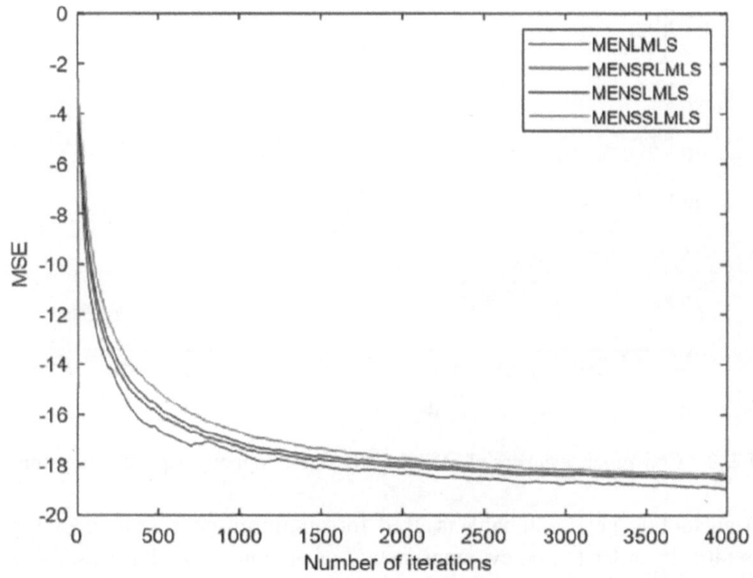

FIGURE 5.7 Convergence curves of MENLMLS and its signed adaptive algorithms.

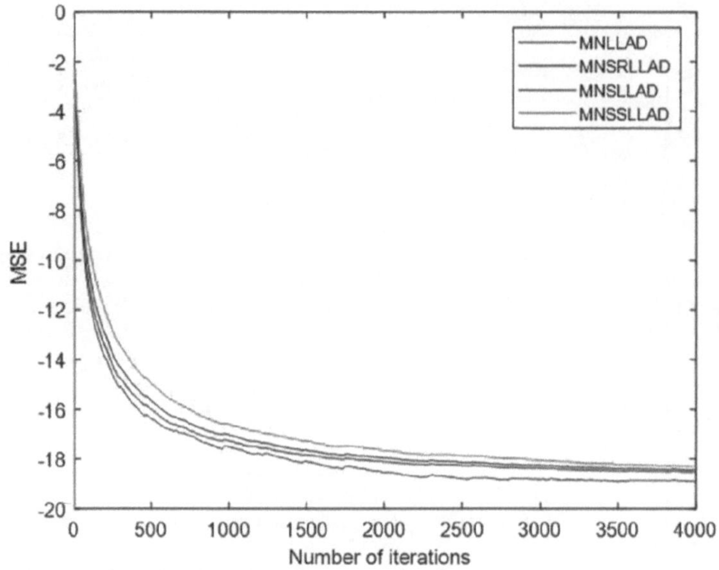

FIGURE 5.8 Convergence curves of MNLLAD and its signed adaptive algorithms.

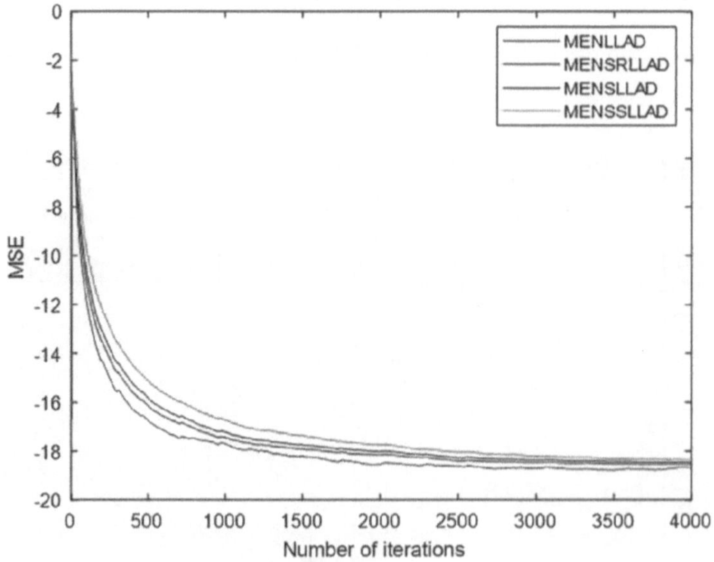

FIGURE 5.9 Convergence curves of MENLLAD and its signed adaptive algorithms.

(NCBI) nucleotide gene databank is used for obtaining gene sequences for DNA analysis [40]. In order to achieve consistency of outcomes, we have used ten DNA sequences of *Homo sapiens* as our dataset for assessing efficiency of multiple adaptive techniques. Description of several sequences taken from the NCBI gene databank is shown in Table 3.8.

5.10.1 Exon Prediction Results

Here, the performance measures of various AEPs developed using maximum normalized logarithmic adaptive algorithms are measured, analyzed, and compared. Various maximum normalized based AEPs are developed using MNLMLS, MENLMLS, MNLLAD, and MENLLAD algorithms along with their sign-based variants. For evaluation of diverse developed AEPs, exon prediction results are used for ten gene sequences from the NCBI gene database. The sequence 5 performance measures with accession AF009962 are assessed using MATLAB for all NLMLS and MNLMLS algorithms and their sign-based variants as shown in Tables 5.7–5.10.

Metrics like Sn, Sp, and Pr are determined by threshold values between 0.4 and 0.9. At a threshold of 0.8, exon identification seems better and the values of

TABLE 5.7
Performance Measures of NLMLS- and MNLMLS-Based AEPs for a Gene Sequence with Accession AF009962

Algorithm	Metric	Gene Sequence Serial Number									
		1	2	3	4	5	6	7	8	9	10
LMS	Sn	0.6286	0.6384	0.6457	0.6273	0.6481	0.6162	0.6193	0.6241	0.6268	0.6202
	Sp	0.6435	0.6628	0.6587	0.6405	0.6518	0.6324	0.6529	0.6289	0.6452	0.5965
	Pr	0.5922	0.5894	0.5934	0.5858	0.5904	0.5786	0.5896	0.5856	0.5814	0.5761
NLMLS	Sn	0.8168	0.7824	0.8128	0.8145	0.8062	0.8145	0.8024	0.8182	0.8024	0.8123
	Sp	0.8065	0.7932	0.8021	0.8024	0.8137	0.8034	0.8145	0.8036	0.8137	0.8021
	Pr	0.8185	0.8036	0.8034	0.8137	0.8085	0.8128	0.8043	0.8154	0.8034	0.8134
NSRLMLS	Sn	0.7993	0.7635	0.7982	0.7936	0.7868	0.7941	0.7824	0.7932	0.7835	0.7972
	Sp	0.7874	0.7741	0.7836	0.7835	0.7945	0.7824	0.7936	0.7845	0.7941	0.7836
	Pr	0.7846	0.7824	0.7923	0.7941	0.7882	0.7924	0.7826	0.7928	0.7824	0.7923
NSLMLS	Sn	0.7806	0.7569	0.7793	0.7735	0.7737	0.7736	0.7687	0.7782	0.7689	0.7727
	Sp	0.7764	0.7575	0.7782	0.7789	0.7798	0.7615	0.7735	0.7654	0.7675	0.7682
	Pr	0.7685	0.7715	0.7767	0.7775	0.7796	0.7784	0.7678	0.7772	0.7715	0.7767
NSSLMLS	Sn	0.7615	0.7297	0.7581	0.7557	0.7554	0.7586	0.7492	0.7562	0.7497	0.7581
	Sp	0.7487	0.7386	0.7965	0.7497	0.7582	0.7526	0.7557	0.7481	0.7486	0.7465
	Pr	0.7592	0.7426	0.7588	0.7586	0.7584	0.7531	0.7462	0.7525	0.7426	0.7588
MNLMLS	Sn	0.8081	0.7768	0.8072	0.8067	0.7987	0.8073	0.7962	0.8109	0.7981	0.8094
	Sp	0.8001	0.7893	0.7999	0.7987	0.8091	0.7984	0.8078	0.7963	0.8092	0.7966
	Pr	0.8098	0.7967	0.7991	0.8098	0.8001	0.8094	0.7998	0.8094	0.7989	0.8082
MNSRLMLS	Sn	0.7901	0.7553	0.7915	0.7891	0.7789	0.7899	0.7787	0.7890	0.7773	0.7894
	Sp	0.7799	0.7679	0.7778	0.7785	0.7890	0.7789	0.7872	0.7785	0.7869	0.7780
	Pr	0.7782	0.7792	0.7882	0.7890	0.7806	0.7887	0.7769	0.7887	0.7791	0.7897
MNSLMLS	Sn	0.7792	0.7486	0.7705	0.7679	0.7685	0.7693	0.7602	0.7701	0.7600	0.7656
	Sp	0.7687	0.7477	0.7715	0.7703	0.7707	0.7543	0.7682	0.7593	0.7599	0.7569
	Pr	0.7598	0.7665	0.7699	0.7684	0.7701	0.7711	0.7606	0.7665	0.7651	0.7723
MNSSLMLS	Sn	0.7579	0.7201	0.7524	0.7488	0.7476	0.7526	0.7400	0.7487	0.7417	0.7535
	Sp	0.7487	0.7312	0.7886	0.7410	0.7505	0.7479	0.7489	0.7407	0.7403	0.7348
	Pr	0.7509	0.7382	0.7516	0.7511	0.7499	0.7484	0.7381	0.7495	0.7385	0.7509

TABLE 5.8
Performance Measures of ENLMLS- and MENLMLS-Based AEPs for a Gene Sequence with Accession AF009962

		Gene Sequence Serial Number									
Algorithm	**Metric**	**1**	**2**	**3**	**4**	**5**	**6**	**7**	**8**	**9**	**10**
LMS	Sn	0.6286	0.6384	0.6457	0.6273	0.6481	0.6162	0.6193	0.6241	0.6268	0.6202
	Sp	0.6435	0.6628	0.6587	0.6405	0.6518	0.6324	0.6529	0.6289	0.6452	0.5965
	Pr	0.5922	0.5894	0.5934	0.5858	0.5904	0.5786	0.5896	0.5856	0.5814	0.5761
ENLMLS	Sn	0.8016	0.7782	0.8032	0.8024	0.8006	0.8014	0.7996	0.8088	0.7985	0.8062
	Sp	0.8006	0.7893	0.8012	0.7994	0.8023	0.7963	0.8084	0.7988	0.8036	0.7981
	Pr	0.8028	0.8003	0.7996	0.8013	0.8008	0.8022	0.7985	0.8035	0.7983	0.8014
ENSRLMLS	Sn	0.7899	0.7563	0.7898	0.7873	0.7786	0.7864	0.7782	0.7893	0.7785	0.7877
	Sp	0.7787	0.7674	0.7783	0.7783	0.7894	0.7782	0.7895	0.7789	0.7894	0.7754
	Pr	0.7784	0.7792	0.7871	0.7894	0.7798	0.7892	0.7782	0.7854	0.7782	0.7857
ENSLMLS	Sn	0.7780	0.7456	0.7679	0.7653	0.7668	0.7673	0.7568	0.7659	0.7572	0.7644
	Sp	0.7673	0.7457	0.7678	0.7664	0.7645	0.7561	0.7693	0.7565	0.7578	0.7582
	Pr	0.7568	0.7671	0.7652	0.7673	0.7679	0.7649	0.7587	0.7692	0.7633	0.7676
ENSSLMLS	Sn	0.7561	0.7129	0.7458	0.7422	0.7455	0.7478	0.7349	0.7468	0.7349	0.7454
	Sp	0.7348	0.7235	0.7896	0.7349	0.7468	0.7452	0.7471	0.7343	0.7387	0.7336
	Pr	0.7459	0.7342	0.7483	0.7474	0.7457	0.7476	0.7356	0.7484	0.7385	0.7466
MENLMLS	Sn	0.7914	0.7675	0.7912	0.7921	0.7900	0.7903	0.7880	0.7982	0.7873	0.7978
	Sp	0.7903	0.7789	0.7901	0.7877	0.7905	0.7857	0.7971	0.7875	0.7909	0.7856
	Pr	0.7921	0.7901	0.7888	0.7907	0.7917	0.7919	0.7874	0.7912	0.7890	0.7900
MENSRLMLS	Sn	0.7789	0.7451	0.7777	0.7781	0.7678	0.7749	0.7650	0.7788	0.7677	0.7753
	Sp	0.7672	0.7566	0.7684	0.7879	0.7785	0.7665	0.7791	0.7676	0.7783	0.7675
	Pr	0.7674	0.7703	0.7763	0.7686	0.7712	0.7787	0.7677	0.7743	0.7671	0.7719
MENSLMLS	Sn	0.7668	0.7343	0.7571	0.7547	0.7559	0.7545	0.7451	0.7560	0.7467	0.7521
	Sp	0.7571	0.7330	0.7559	0.7561	0.7540	0.7450	0.7591	0.7458	0.7471	0.7479
	Pr	0.7462	0.7593	0.7546	0.7536	0.7561	0.7534	0.7470	0.7588	0.7502	0.7551
MENSSLMLS	Sn	0.7456	0.7015	0.7335	0.7301	0.7322	0.7366	0.7236	0.7349	0.7230	0.7335
	Sp	0.7237	0.7119	0.7791	0.7230	0.7354	0.7349	0.7363	0.7212	0.7277	0.7249
	Pr	0.7361	0.7202	0.7379	0.7359	0.7348	0.7361	0.7250	0.7366	0.7264	0.7351

performance measures using NLMLS- and MNLMLS-based algorithms along with their signed versions are as shown in Table 5.7.

From the values, it is evident that the performance measures of NLMLS- and MNLMLS-based AEPs are just inferior compared to NSRLMLS- and MNSRLMLS-based AEPs. For a larger number of iterations, the computational complexity of NLMLS and MNLMLS algorithms is higher than NSRLMLS and MNSRLMLS algorithms. In such cases, MNSRLMLS algorithm becomes efficient due to its significantly low complexity in performing the computations and in terms of exon locating ability. The signum function present in all signed versions of MNLMLS reduces the computational complexity, and thus, all signed versions predict the exon locations more accurately. Of all these algorithms, MNSRLMLS-based AEP is

TABLE 5.9
Performance Measures of NLLAD- and MNLLAD-Based AEPs for a Gene Sequence with Accession AF009962

		Gene Sequence Serial Number									
Algorithm	Metric	1	2	3	4	5	6	7	8	9	10
LMS	Sn	0.6286	0.6384	0.6457	0.6273	0.6481	0.6162	0.6193	0.6241	0.6268	0.6202
	Sp	0.6435	0.6628	0.6587	0.6405	0.6518	0.6324	0.6529	0.6289	0.6452	0.5965
	Pr	0.5922	0.5894	0.5934	0.5858	0.5904	0.5786	0.5896	0.5856	0.5814	0.5761
NLLAD	Sn	0.8523	0.8553	0.8534	0.8618	0.8537	0.8445	0.8532	0.8582	0.8548	0.8572
	Sp	0.8424	0.8462	0.8464	0.8514	0.8468	0.8575	0.8542	0.8476	0.8472	0.8486
	Pr	0.8506	0.8531	0.8561	0.8538	0.8578	0.8643	0.8531	0.8586	0.8565	0.8593
NSRLLAD	Sn	0.8012	0.8102	0.8079	0.8094	0.8037	0.8125	0.8142	0.8054	0.8075	0.8054
	Sp	0.8124	0.8164	0.8114	0.8138	0.8138	0.8086	0.8074	0.8106	0.8154	0.8187
	Pr	0.8096	0.8196	0.8097	0.8068	0.8045	0.8102	0.8029	0.8124	0.8086	0.8076
NSLLAD	Sn	0.7862	0.7911	0.7891	0.7868	0.7878	0.7954	0.7988	0.7882	0.7882	0.7892
	Sp	0.7902	0.7895	0.7936	0.7995	0.7931	0.7878	0.7874	0.7976	0.7954	0.7964
	Pr	0.7896	0.7934	0.7892	0.7894	0.7892	0.7872	0.7868	0.7862	0.7895	0.7914
NSSLLAD	Sn	0.7694	0.7708	0.7686	0.7697	0.7705	0.7596	0.7546	0.7732	0.7715	0.7718
	Sp	0.7586	0.7611	0.7592	0.7535	0.7587	0.7632	0.7632	0.7594	0.7592	0.7592
	Pr	0.7612	0.7586	0.7618	0.7697	0.7612	0.7704	0.7714	0.7676	0.7635	0.7684
MNLLAD	Sn	0.8472	0.8465	0.8453	0.8561	0.8443	0.8364	0.8453	0.8441	0.8465	0.8457
	Sp	0.8342	0.8346	0.8367	0.8454	0.8346	0.8457	0.8447	0.8374	0.8347	0.8358
	Pr	0.8450	0.8477	0.8468	0.8483	0.8457	0.8564	0.8423	0.8488	0.8456	0.8489
MNSRLLAD	Sn	0.7988	0.8010	0.8007	0.8039	0.7992	0.8083	0.8094	0.8005	0.8027	0.8015
	Sp	0.8012	0.8016	0.8021	0.8083	0.8075	0.8006	0.8027	0.8082	0.8115	0.8108
	Pr	0.8029	0.8117	0.8013	0.7996	0.8004	0.8010	0.7988	0.8076	0.8008	0.8017
MNSLLAD	Sn	0.7786	0.7891	0.7764	0.7754	0.7787	0.7895	0.7894	0.7781	0.7753	0.7789
	Sp	0.7890	0.7789	0.7893	0.7879	0.7835	0.7762	0.7743	0.7863	0.7894	0.7856
	Pr	0.7789	0.7893	0.7745	0.7782	0.7761	0.7749	0.7786	0.7754	0.7789	0.7891
MNSSLLAD	Sn	0.7565	0.7670	0.7578	0.7549	0.7670	0.7457	0.7434	0.7677	0.7631	0.7670
	Sp	0.7458	0.7561	0.7449	0.7433	0.7478	0.7563	0.7563	0.7454	0.7472	0.7485
	Pr	0.7561	0.7458	0.7561	0.7574	0.7588	0.7670	0.7657	0.7565	0.7593	0.7568

effective in terms of accurate exon prediction when compared to LMS, NLMLS, and its other signed variants using sensitivity Sn 0.7789 (77.89%), specificity Sp 0.7890 (78.90%), and precision Pr 0.7806 (78.06%), respectively, which are a bit inferior than MNLMLS- and MENLMLS-based AEPs. But, due to its significantly low computational complexity, MNSRLMS-based AEP outperforms all other NLMLS- and MNLMLS-based adaptive variants.

Locating exon segments would be better at threshold 0.8 for SRA-based AEP for a gene sequence with accession AF009962. A typical PSD plot for MNLMLS and its sign-based variants is shown in Figure 5.10. It shows the predicted exon location of sequence 5 applying LMS, MNLMLS, and its sign-based techniques. From this plot, it is evident that AEP using LMS is not correctly located exon sections. It creates

TABLE 5.10
Performance Measures of ENLLAD- and MENLLAD-Based AEPs for a Gene Sequence with Accession AF009962

		Gene Sequence Serial Number									
Algorithm	**Metric**	**1**	**2**	**3**	**4**	**5**	**6**	**7**	**8**	**9**	**10**
LMS	Sn	0.6286	0.6384	0.6457	0.6273	0.6481	0.6162	0.6193	0.6241	0.6268	0.6202
	Sp	0.6435	0.6628	0.6587	0.6405	0.6518	0.6324	0.6529	0.6289	0.6452	0.5965
	Pr	0.5922	0.5894	0.5934	0.5858	0.5904	0.5786	0.5896	0.5856	0.5814	0.5761
ENLLAD	Sn	0.8296	0.8292	0.8285	0.8238	0.8282	0.8263	0.8288	0.8265	0.8294	0.8294
	Sp	0.8235	0.8273	0.8274	0.8297	0.8246	0.8323	0.8365	0.8252	0.8262	0.8265
	Pr	0.8311	0.8328	0.8335	0.8378	0.8311	0.8245	0.8302	0.8347	0.8378	0.8342
ENSRLLAD	Sn	0.7732	0.7724	0.7795	0.7736	0.7734	0.7846	0.7882	0.7751	0.7756	0.7758
	Sp	0.7811	0.7863	0.7835	0.7876	0.7825	0.7752	0.7792	0.7848	0.7875	0.7864
	Pr	0.7796	0.7785	0.7737	0.7778	0.7783	0.7765	0.7746	0.7783	0.7792	0.7808
ENSLLAD	Sn	0.7654	0.7642	0.7688	0.7685	0.7651	0.7658	0.7702	0.7672	0.7656	0.7698
	Sp	0.7695	0.7645	0.7644	0.7697	0.7685	0.7746	0.7754	0.7696	0.7684	0.7686
	Pr	0.7705	0.7783	0.7736	0.7736	0.7718	0.7682	0.7696	0.7724	0.7735	0.7712
ENSSLLAD	Sn	0.7503	0.7552	0.7595	0.7545	0.7522	0.7582	0.7572	0.7528	0.7542	0.7545
	Sp	0.7538	0.7567	0.7576	0.7564	0.7554	0.7598	0.7602	0.7562	0.7584	0.7568
	Pr	0.7592	0.7525	0.7593	0.7582	0.7608	0.7562	0.7714	0.7624	0.7627	0.7674
MENLLAD	Sn	0.8129	0.8119	0.8148	0.8173	0.8188	0.8013	0.8026	0.8125	0.8149	0.8156
	Sp	0.8143	0.8137	0.8166	0.8159	0.8124	0.8024	0.8124	0.8168	0.8126	0.8184
	Pr	0.8231	0.8252	0.8233	0.8237	0.8241	0.8035	0.8150	0.8274	0.8277	0.8265
MENSRLLAD	Sn	0.7834	0.7845	0.7914	0.7826	0.7812	0.7836	0.7845	0.7824	0.7902	0.7858
	Sp	0.7782	0.7816	0.7841	0.7765	0.7836	0.7821	0.7763	0.7816	0.7871	0.7787
	Pr	0.7879	0.7873	0.7857	0.7879	0.7867	0.7795	0.7854	0.7857	0.7829	0.7839
MENSLLAD	Sn	0.7565	0.7642	0.7568	0.7522	0.7585	0.7565	0.7571	0.7567	0.7565	0.7549
	Sp	0.7583	0.7564	0.7544	0.7539	0.7568	0.7674	0.7675	0.7559	0.7558	0.7574
	Pr	0.7670	0.7678	0.7673	0.7651	0.7671	0.7568	0.7549	0.7672	0.7673	0.7671
MENSSLLAD	Sn	0.7450	0.7465	0.7479	0.7484	0.7442	0.7456	0.7437	0.7452	0.7454	0.7454
	Sp	0.7463	0.7476	0.7433	0.7467	0.7455	0.7449	0.7560	0.7476	0.7438	0.7463
	Pr	0.7439	0.7442	0.7459	0.7478	0.7560	0.7457	0.7671	0.7534	0.7582	0.7577

certain ambiguity to locate position in the form of a short peak for exon location and some unwanted peaks at different positions in the power spectrum.

For signed variants based on MNLMLS, MNSRLMLS algorithm located the exon position at 3934-4581 with excellent intensity, and also a sharp peak in PSD plot is observed. The PSD plots of MNLMLS and its signed versions are shown in Figure 5.10b–e. Due to sign function, exon locating ability of these techniques remains better compared to LMS. In the event of MNSSLMLS, because of input clipped sequence and feedback signal, exon locating ability is lower than its other signed variants.

Based on small computational difficulty, performance measures in Table 5.7, and exon prediction plots in Figure 5.10, it is obvious that AEP using MNSRLMLS is

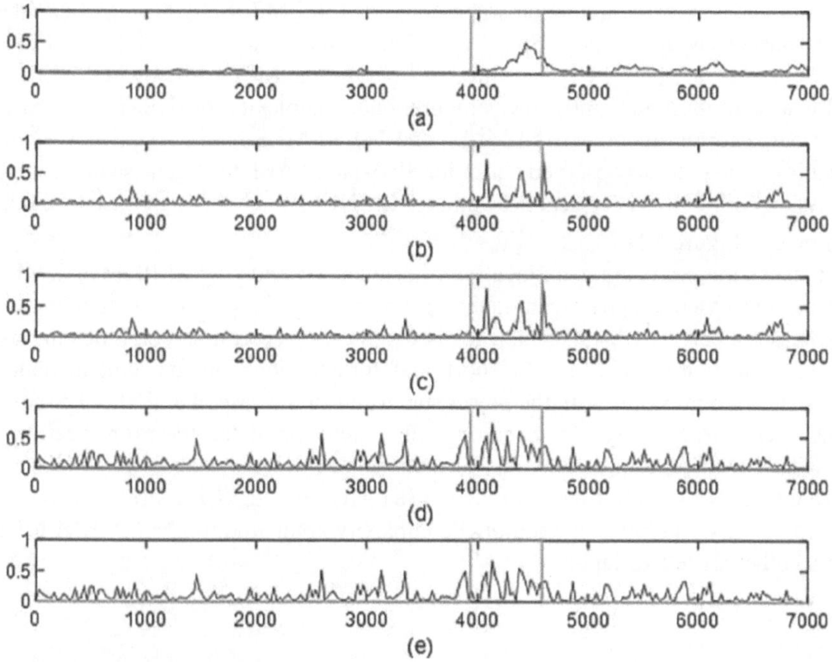

FIGURE 5.10 PSD plots with the location of exon (3934-4581) for a DNA sequence with accession AF009962 predicted using (a) LMS-based AEP, (b) MNLMLS-based AEP, (c) MNSRLMLS-based AEP, (d) MNSLMLS-based AEP, and (e) MNSSLMLS based AEP. (Relative base location is taken on x-axis, and power spectrum is taken on y-axis.)

the best candidate in applications relevant to exon identification. In Figure 5.10a, a relatively short peak is evident for exon location, and certain undesirable peaks are located at different positions in the power spectrum plot using LMS algorithm. Simultaneously, actual exon location 3934-4581 is not predicted accurately using LMS algorithm.

The values of prediction metrics using SRLMS, SLMS, and SSLMS techniques from Table 3.4 are observed as inferior to all MNLMLS-based adaptive algorithms. Similarly, the performance measures of sequence 5 with accession AF009962 for all ENLMLS- and MENLMLS-based algorithms along with their signed versions are as shown in Table 5.8. From these values, it is evident that the performance measures of ENLMLS- and MENLMLS-based AEPs are just inferior compared to ENSRLMLS- and MENSRLMLS-based AEPs. For a larger number of iterations, the computational complexity of ENLMLS and MENLMLS algorithms is higher than ENSRLMLS and MENSRLMLS algorithms. In such cases, MENSRLMLS algorithm becomes efficient due to its significantly low complexity in performing the computations and in terms of exon locating ability.

The signum function present in all signed versions of MENLMLS reduces the computational complexity, and they have predicted the exon locations more accurately compared to LMS. Of all these algorithms, MENSRLMLS-based AEP is effective in terms of accurate exon prediction when compared to LMS, ENLMLS, and its other signed variants with specificity Sp 0.7785 (77.85%), sensitivity Sn 0.7678 (76.78%), and precision Pr 0.7712 (77.12%), which are just inferior than MENLMLS-based AEP.

Because of its significantly low computational complexity, MENSRLMLS-based AEP outperforms all other ENLMLS- and MENLMLS-based adaptive variants. Locating exon sections observed better for SRA-based AEP for a gene sequence with accession AF009962. A typical PSD plot for MENLMLS and its sign-based versions is shown in Figure 5.11.

It shows the predicted exon location of sequence 5 applying LMS, MENLMLS, and its sign-based adaptive techniques. From this plot, it is evident that AEP with LMS has not correctly located exon segments. It creates certain ambiguities in position identification in the form of a short peak for exon location and some unwanted peaks at different positions in the power spectrum. In the case of MENLMLS-based signed versions, MENSRLMLS-based AEP exactly predicted the exon location at 3934-4581 with good intensity, and a sharp peak in PSD plot is observed.

Due to use of signum function in MENSRLMLS-based AEP, it offers good exon location ability and low computational complexity compared to LMS, MENLMLS, and its other signed variants.

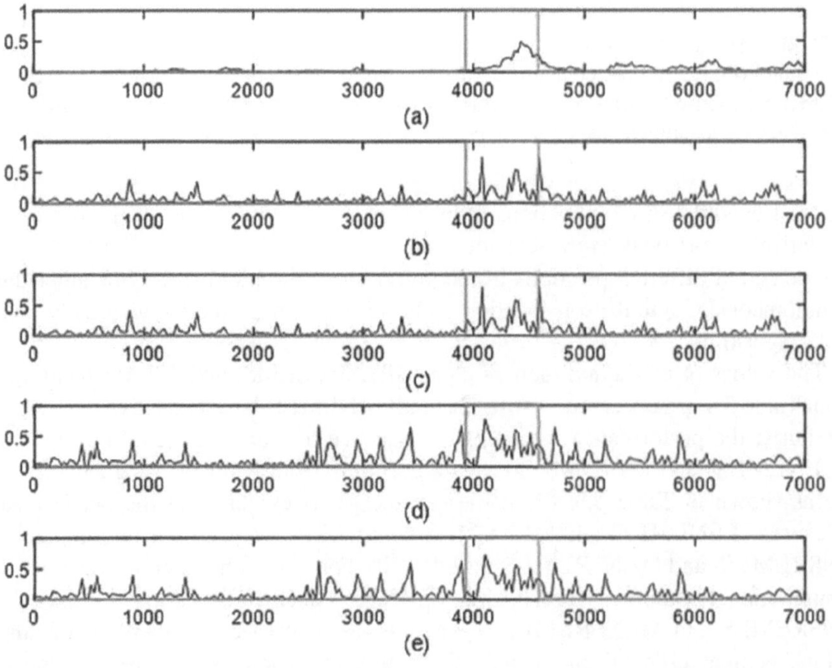

FIGURE 5.11 PSD plots with location of exon (3934-4581) for a DNA sequence with accession AF009962 predicted using (a) LMS-based AEP, (b) MENLMLS-based AEP, (c) MENSRLMLS-based AEP, (d) MENSLMLS-based AEP, and (e) MENSSLMLS-based AEP. (Relative base location is taken on x-axis, and power spectrum is taken on y-axis.)

The PSD plots for MENLMLS and its signed variants are shown in Figure 5.11b–e. Because of the sign function involved in these algorithms, tracking capability of these algorithms is better than LMS algorithm. In the event of MENSSLMLS, owing to clipped input sequence and clipped feedback signal, performance of exon identification is lower than its other signed versions. Thus, from low computational complexity, performance measures in Table 5.8, and exon prediction plots in Figure 5.11, it is shown that MENSRLMLS-based AEP is found to be the best candidate in exon identification applications.

In Figure 5.11a, a relatively short peak is evident for exon location and some unwanted peaks at distinct places in the PSD plot using LMS algorithm are recognized. Actual exon location 3934-4581 is not identified at the same time accurately using LMS algorithm. It is clear that MENSRLMLS-based AEP predicts the exon location with a sharp peak accurately compared to all other MENLMLS variants. The values of prediction measures using SRLMS, SLMS, and SSLMS algorithms from Table 3.4 are observed as inferior to all MENLMLS-based adaptive algorithms. Due to use of signum function in MENSRLMLS-based AEP, it offers good exon location ability as well as low computing difficulty compared to LMS, MENLMLS, and its other signed variants.

Likewise, the performance measures of sequence 5 with accession AF009962 for all NLLAD- and MNLLAD-based algorithms along with their signed versions are as shown in Table 5.9. From these values, it is evident that the performance measures of NLLAD- and MNLLAD-based AEPs are just inferior compared to NSRLLAD- and MNSRLLAD-based AEPs. For a larger number of iterations, the computational complexity of NLLAD and MNLLAD algorithms becomes more than NSRLLAD and MNSRLLAD algorithms. In such cases, MNSRLLAD algorithm becomes efficient due to its significantly low complexity in performing the computations and in terms of exon locating ability.

The signum function present in all signed versions of MNLLAD reduces the computational complexity, and they have predicted the exon locations more accurately compared to LMS. Of all these algorithms, MNSRLLAD-based AEP is effective in terms of accurate exon prediction when compared to LMS, MNLLAD, and its other signed variants using sensitivity Sn 0.7992 (79.92%), specificity Sp 0.8075 (80.75%), and Precision Pr 0.8004 (80.04%) respectively, which are just inferior than MNLLAD-based AEP. But, because of its significantly low computational complexity, MNSRLLAD-based AEP outperforms all other NLLAD- and MNLLAD-based adaptive variants.

A typical PSD plot for MNLLAD and its sign-based variants is shown in Figure 5.12. It shows the predicted exon location of sequence 5 applying LMS, MNLLAD, and its signed adaptive techniques. It is obvious that AEP using LMS has not correctly located exon position.

It causes some ambiguities in location identification in the form of a short peak for exon location and some unwanted peaks at different positions in the power spectrum. In the case of MNLLAD-based signed versions, MNSRLLAD-based AEP exactly predicts the exon location at 3934-4581 with good intensity and a sharp peak in PSD plot is observed. Due to use of signum function in MNSRLLAD-based AEP, it offers good exon location ability and low computational complexity compared to LMS, MNLLAD, and its other signed variants.

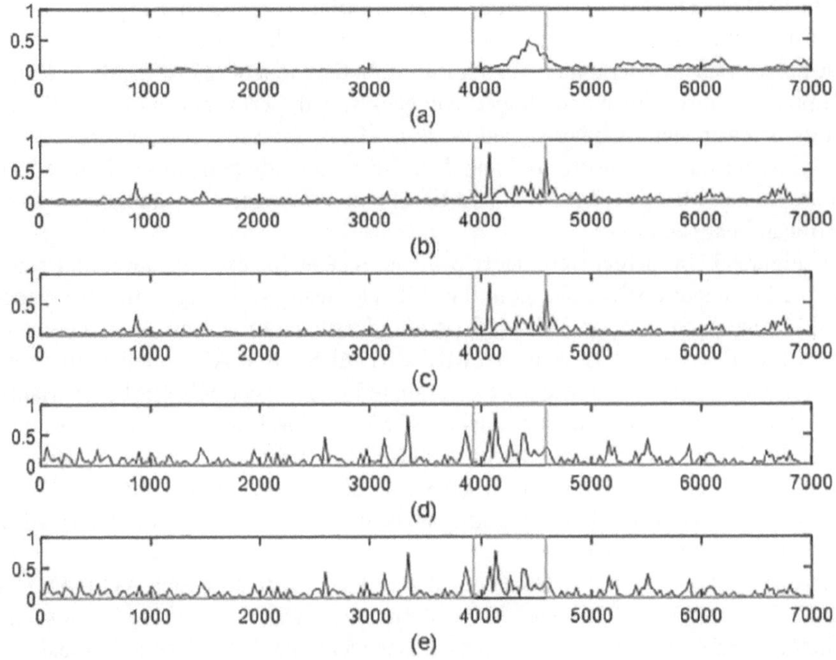

FIGURE 5.12 PSD plots with the location of exon (3934-4581) for a DNA sequence with accession AF009962 predicted using (a) LMS-based AEP, (b) MNLLAD-based AEP, (c) MNSRLLAD-based AEP, (d) MNSLLAD-based AEP, and (e) MNSSLLAD-based AEP. (Relative base location is taken on x-axis, and power spectrum is taken on y-axis.)

The PSD plots for MNLLAD and its signed variants are shown in Figure 5.12b–e. The exon locating capability of these algorithms is better than LMS algorithm because of its sign function.

In the case of MNSSLLAD, performance in terms of exon identification is lower compared to other signed variants because of clipped sequence of input and clipped feedback signal. In Figure 5.12a, a relatively short peak is evident for exon location and certain unwanted peaks are recognized at different locations in PSD plot using LMS algorithm. Therefore, based on low computational complexity, performance measures in Table 5.9, and exon prediction plots in Figure 5.12, it is found that MNSRLLAD-based AEP is found to be the better candidate in exon prediction applications.

At the same time, the actual exon location 3934-4581 is not predicted accurately using LMS algorithm. It is clear that MNSRLLAD-based AEP predicted the exon location with a sharp peak accurately compared to all other MNLLAD variants. The values of prediction measures such as sensitivity, specificity, and precision of SRLMS, SLMS, and SSLMS algorithms from Table 3.4 are observed as inferior to all MNLLAD and its sign-based adaptive algorithms. Due to the use of signum function in MNSRLLAD-based AEP, it offers good exon location ability and low computational complexity compared to LMS, MNLLAD, and its other signed variants.

Also, the performance measures of sequence 5 with accession AF009962 for all ENLLAD- and MENLLAD-based algorithms along with their signed versions are as shown in Table 5.10. From these values, it is evident that the performance measures of ENLLAD- and MENLLAD-based AEPs are just inferior to ENSRLLAD- and MENSRLLAD-based AEPs. For a larger number of iterations, the computational complexity of ENLLAD and MENLLAD algorithms becomes more than ENSRLLAD and MENSRLLAD algorithms.

In such cases, MENSRLLAD algorithm becomes efficient due to its significantly low complexity in performing the computations and in terms of exon locating ability. The signum function present in all signed versions of MENLLAD reduces computational complexity, and they have predicted the exon locations more accurately compared to LMS. Of all these algorithms, MENSRLLAD-based AEP is effective in terms of accurate exon prediction when compared to LMS, MENLLAD, and its other signed variants with sensitivity Sn 0.7812 (78.12%), specificity Sp 0.7836 (78.36%), and precision Pr 0.7867 (78.67%), respectively, which are just inferior than MENLLAD-based AEP.

But, due to its significantly low computational complexity, MENSRLLAD-based AEP outperforms all other NLLAD- and MENLLAD-based adaptive variants. At threshold 0.8, exon prediction seems to be better for sign regressor–based AEP for a gene sequence with accession AF009962.

A typical PSD plot for MENLLAD and its sign-based variants is shown in Figure 5.13. It shows the predicted exon location of sequence 5 applying LMS, MENLLAD, and its signed adaptive algorithms. From this plots, it is clear that LMS-based AEP has not accurately located the exon position.

This algorithm causes some ambiguities in location prediction in the form of a short peak for exon location and some unwanted peaks at different positions in the power spectrum. In the case of MENLLAD-based signed versions, MENSRLLAD-based AEP exactly predicts the exon location at 3934-4581 with good intensity, and a sharp peak in PSD plot is observed. Due to the use of signum function in MENSRLLAD-based AEP, it offers good exon location ability and low computational complexity compared to LMS, MENLLAD, and its other signed variants. The PSD plots for MENLLAD and its signed variants are shown in Figure 5.13b–e. Because of the sign function involved in these algorithms, the tracking capability of these algorithms is better than LMS algorithm. In the case of MENSSLLAD, performance of exon prediction is lower compared to its other signed versions, because of its clipped input sequence and clipped feedback signal.

Therefore, based on low computational complexity, performance measures in Table 5.6, and exon prediction plots in Figure 5.13, it is found that MENSRLLAD-based AEP is found to be the better candidate in exon prediction applications. In Figure 5.13a, a relatively short peak is evident for exon location and certain undesirable peaks are evident at different positions in PSD plot using LMS algorithm. At the same moment, the actual exon location 3934-4581 is not predicted accurately using LMS algorithm. It was evident that AEP based on MVNSRLMS located the exon position correctly with a sharp peak compared to all other MVNLMS variants. The values of prediction measures such as sensitivity, specificity, and precision of SRLMS, SLMS, and SSLMS algorithms from Table 3.4 are observed as inferior to

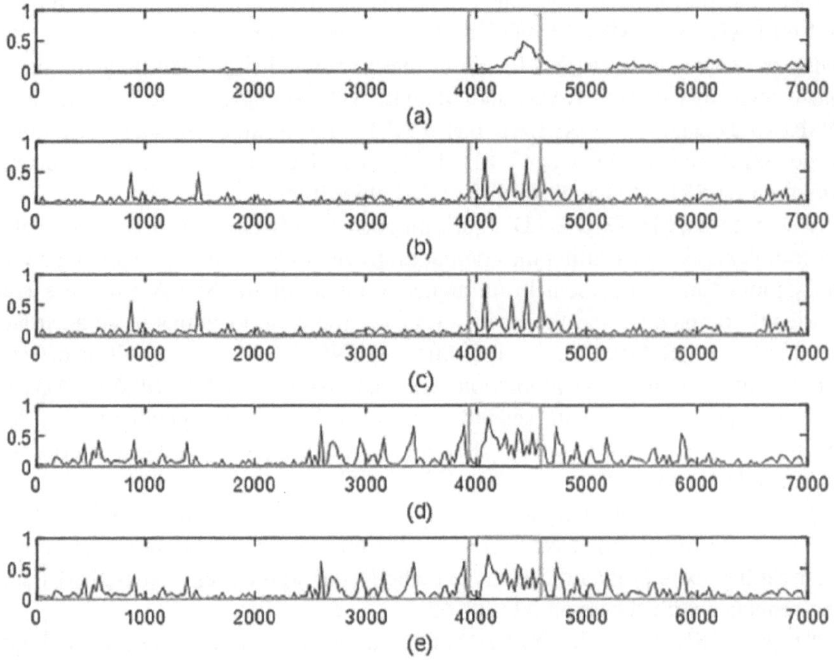

FIGURE 5.13 PSD plots with location of exon (3934-4581) for a DNA sequence with accession AF009962 predicted using (a) LMS-based AEP, (b) MENLLAD-based AEP, (c) MENSRLLAD-based AEP, (d) MENSLLAD-based AEP, and (e) MENSSLLAD-based AEP. (Relative base location is taken on x-axis, and power spectrum is taken on y-axis.)

all MENLLAD-based adaptive algorithms. Due to the use of signum function in MENSRLLAD-based AEP, it offers good exon location ability and low computational complexity compared to LMS, MNLMF, and its other signed variants.

5.11 CONCLUSIONS

In this chapter, we have developed weight update variants of normalization- and maximum normalization-based realization of various logarithmic adaptive algorithms. Distinct AEPs are created using these algorithms for accurate prediction of exon locations in real gene sequences of *Homo sapiens*. Varieties of AEPs are developed and examined using eight normalization-based adaptive algorithms and their sign-based variants on original DNA datasets of *Homo sapiens* taken from the NCBI gene databank. An introduction to logarithmic adaptive algorithms, an account of detailed explanation about different logarithmic normalized adaptive algorithms including NLMLS, ENLMLS, NLLAD, and ENLLAD, and an extension to their sign-based realizations are presented. The LMS algorithm is considered as the reference algorithm, and various logarithmic realizations based on normalization are discussed.

To evaluate the performance of various AEPs, we have plotted convergence characteristics, PSD plots, also computational complexities, and measures like sensitivity

(Sn), specificity (Sp), and precision (Pr) in the results and discussion section followed by conclusions. Here, we have inspected the possibility of enhancing the existing methods in order to achieve better performance. In order to accomplish this, it is desirable to reduce computational complexity and improve the exon locating ability. This can be achieved by either clipping of input data or estimation error. Three signed algorithms are derived based on this clipping, and we have extended the NLMLS-based adaptive algorithms to three versions of sign-based adaptive algorithms. To further reduce the computational complexity significantly and provide better performance, we have extended all the normalized based logarithmic adaptive algorithms to maximum normalized based logarithmic adaptive algorithms. This is the third category of the adaptive algorithms we have considered in this chapter.

Totally, thirty-two AEPs using logarithmic normalized adaptive algorithms are developed and tested on ten real gene sequences of *Homo sapiens* taken from the NCBI databank. Of these, sixteen normalized adaptive algorithms along with their sign-based algorithms are used for development of various AEPs including NLMLS, NSRLMLS, NSLMLS, NSSLMLS, ENLMLS, ENSRLMLS, ENSLMLS, ENSSLMLS, NLLAD, NSRLLAD, NSLLAD, NSSLLAD, ENLLAD, ENSRLLAD, ENSLLAD, and ENSSLLAD algorithms. Among all the sixteen normalized logarithmic adaptive algorithms, NSRLLAD-based AEP performed better in terms of exon tracking ability, low computational complexity, and convergence performance.

Similarly, various AEPs are developed using sixteen maximum normalization-based realization of logarithmic adaptive algorithms and their sign-based algorithms including MNLMLS, MNSRLMLS, MNSLMLS, MNSSLMLS, MENLMLS, MENSRLMLS, MENSLMLS, MENSSLMLS, MNLLAD, MNSRLLAD, MNSLLAD, MNSSLLAD, MENLLAD, MENSRLLAD, MENSLLAD, and MENSSLLAD algorithms. Among all these sixteen maximum normalized adaptive algorithms, MNSRLLAD-based AEP performs better in terms of exon tracking ability, low computational complexity, and better convergence performance than LMS algorithm and other variants based on MNLLAD.

Among the four logarithmic adaptive algorithms based on normalization, MNLLAD algorithm performs better compared to all other normalized adaptive algorithms in terms of exon tracking ability, low computational complexity, and convergence performance. Among all the sign-based normalized and maximum normalized logarithmic adaptive algorithms, all their sign regressor versions performed better. All their SA- and SSA-based AEPs are slightly inferior to their LMS counterpart. Overall, all their logarithmic sign-based AEPs deliver greater performance than existing LMS pertaining to computational complexity and metrics like sensitivity, precision, and specificity attained using a gene sequence with accession AF009962 at a threshold value of 0.8. Among all the normalization- and maximum normalization-based AEPs, it is obvious that MNSRLLAD-based AEP is more efficient in terms of exon prediction accuracy, performance measures, and convergence efficiency due to its low computational complexity. The SRNLLAD-based AEP therefore seems to be better than all other adaptive algorithms in all three classifications. Hence, NSRLLAD and MNSRLLAD AEPs could be used in developing low-complexity nano-bioinformatics devices in SOC and LOC applications.

6 Conclusion and Future Perspective

6.1 SUMMARY AND CONCLUSIONS

In Chapter 3, it was shown that all the developed AEPs have accurately predicted the exon location in real DNA sequences of *Homo sapiens*. So as to do this, we have considered sign LMS-based realization of various adaptive algorithms including LMS, namely, least mean fourth (LMF), variable step size LMS (VSLMS), least mean logarithmic squares (LMLS), and least logarithmic absolute difference (LLAD) algorithms and proved that the developed AEPs are more accurate in terms of exon prediction. For this, we considered four measures including PSD plots, computational complexities, convergence performance, and performance metrics sensitivity, specificity, and precision. To accomplish better performance, low computational complexity is needed, and this can be accomplished thru clipping of data or error. The three versions of signed algorithms SRA, SA, and SS make use of this clipping, so we extended the LMS-based adaptive algorithms to make them into sign-based adaptive algorithms. These were used to develop AEPs that were tested on real genomic signals of *Homo sapiens*. To evaluate their performance, we measured computational complexities, performance metrics, and PSD, and convergence characteristics are plotted.

From the results presented in Chapter 3, it is evident that SRA version among signed versions needs a lower number of multiplications. Also, it is evident from the convergence curves that the sign regressor versions of all adaptive algorithms are little bit inferior than its LMS counterparts. From the filtering results, it is little bit inferior than its LMS counterpart. But due to the advantage of sign regressor version that it needs a smaller number of multiplications by an amount equal to filter length, sign regressor version is considered as a better candidate among the considered algorithms for the development of AEPs for accurate exon prediction in DNA sequences.

Among all their sign-based versions, SRLMS, SRLMF, SRVSLMS, SRLMLS, and SRLLAD algorithms perform better with reference to exon locating ability, convergence performance, computational complexity, and performance measures. The performance of LMS and various sign regressor–based AEPs in terms of specificity Sp, sensitivity Sn, and precision Pr is shown in Table 6.1. More the value of all these metrics, the better the exon prediction accuracy for an adaptive algorithm. However, it also depends on the low computational complexity offered by an adaptive algorithm for better performance. From this table, it is obvious that AEP based on SRLLAD better perform for accurate exon prediction than LMS and all other sign regressor–based LMS adaptive techniques due to its low computational complexity.

TABLE 6.1
Performance Measures of Sign Regressor LMS-Based AEPs for a Gene Sequence with Accession AF009962

Gene sequence 5	Metric	LMS	SRLMS	SRLMF	SRVSLMS	SRLMLS	SRLLAD
with accession	Sn	0.6481	0.5849	0.6649	0.6749	0.7565	0.8162
AF009962	Sp	0.6518	0.6105	0.6585	0.6625	0.7692	0.8086
	Pr	0.5904	0.5751	0.6451	0.6751	0.7554	0.8185

In Chapter 4, the problem of improving the accuracy of locating the exons from DNA sequences, different AEPs using various normalization- and maximum normalization-based adaptive algorithms along with their sign-based versions are developed. All these AEPs are tested on ten real genomic sequences of *Home sapiens* obtained from the NCBI gene database. For this, we developed four types of normalized algorithms along with their signed variants based on gradient-based LMS adaptive algorithms discussed in Chapter 3. These include normalized LMS (NLMS), error-normalized LMS (ENLMS), normalized LMF (NLMF), and variable step size normalized LMS (VNLMS) algorithms. To diminish computing difficulty, we successfully developed various AEPs that use sign-based versions of all the four normalized algorithms. Block-based processing using maximum normalization is applied to all the algorithms to further reduce the number of multiplications in divisor of weight update recursion.

Among NLMS and ENLMS algorithms, the performance of NLMS seems better based on exon locating ability, computing difficulty, convergence performance, and performance metrics specificity (Sp), sensitivity (Sn), and precision (Pr). From these two normalized algorithms, we have obtained six sign-based versions of algorithms, and totally eight AEPs are developed for exon prediction. Among the six sign-based normalized AEPs, NSRLMS is in the first place in terms of better performance followed by ENSRLMS, NSLMS, and NSSLMS algorithms in comparison with AEP based on LMS. From the simulation results provided in Chapter 4, it is evident that all the sign regressor versions of maximum normalization–based adaptive algorithms need a smaller number of multiplications than their other counterparts due to the fact that sign regressor algorithm needs lesser multiply computations thru filter length, and block processing using maximum normalization further reduces the number of multiply operations in denominator of weight recursion by a value similar to length of filter.

Hence, among the algorithms presented in Chapter 4, all the sign regressor versions of algorithms are considered as better candidates for accurate exon prediction because of their significantly low computational complexity with respect to exon locating ability, convergence performance, computational complexity, and performance measures. The performance of various sign regressor–based maximum normalized AEPs including LMS base AEP in terms of performance metrics is shown in Table 6.2. More the value of all these metrics, the better the exon prediction accuracy for an adaptive algorithm. However, it also depends on the low computational complexity offered by an adaptive algorithm for better performance. From this table, it

TABLE 6.2
Performance Measures Using Sign Regressor–Based Maximum Normalized AEPs for a Gene Sequence with Accession AF009962

Gene sequence 5	Metric	LMS	MNSRLMS	MENSRLMS	MNSRLMF	MVNSRLMS
with accession	Sn	0.6481	0.6954	0.6785	0.7018	0.7546
AF009962	Sp	0.6518	0.7011	0.6893	0.7278	0.7583
	Pr	0.5904	0.6567	0.6348	0.6869	0.7525

is evident that AEP that relies on MVNSRLMS has shown better performance for accurate exon prediction compared to LMS-based AEP and all other sign regressor–based maximum normalized adaptive techniques because of low computing difficulty. This algorithm also performs better in terms of performance measures than all the sign LMS relevant adaptive techniques used in Chapter 3.

Similarly, in Chapter 5, for improving the accuracy of locating the exons in DNA sequences, different AEPs are developed using various normalization- and maximum normalization-based logarithmic adaptive algorithms along with their sign-based versions. All these AEPs tested on ten real genomic sequences of *Home sapiens* obtained from the NCBI gene database. For this, we developed four types of normalized algorithms along with their signed variants based on gradient-based LMS adaptive algorithms discussed in Chapter 3.

These include normalized LMLS (NLMLS), error-normalized LMLS (ENLMLS), normalized LLAD (NLLAD), and error-normalized LLAD (ENLLAD) algorithms. To reduce computational complexity, we successfully developed various AEPs that use sign-based versions of all the four normalized algorithms. Block-based processing using maximum normalization is applied to all the algorithms to further reduce the number of multiplications in the denominator part of the weight update recursion.

Among NLMLS, ENLMLS, NLLAD, and ENLLAD algorithms, the performance of NLLAD is better depending on exon locating ability, computational complexity, convergence performance, and performance metrics. From these four normalized algorithms, we have obtained six sign-based versions of algorithms, and totally sixteen AEPs are developed for exon prediction. Between the sixteen sign-based normalized AEPs, NSRLLAD is in the first place in terms of better performance in comparison with LMS and other normalization-based logarithmic AEPs. From the simulation results presented in Chapter 5, it is evident that all the sign regressor versions of maximum normalization-based logarithmic adaptive algorithms need a smaller number of multiply computations than their other counterparts due to the fact that sign regressor algorithm needs lesser computations similar to length of filter, and block processing using maximum normalization further reduces the number of multiply calculations in divisor of the weight update recursion thru a value similar to length of filter.

Hence, among the algorithms presented in Chapter 4, all the sign regressor versions of algorithms are considered as better candidates for accurate exon prediction due to their substantially low computational difficulty with regard to locating the

TABLE 6.3
Performance Measures Using Sign Regressor–Based Maximum Normalized Logarithmic AEPs for a Gene Sequence with Accession AF009962

Gene sequence 5	Metric	LMS	MNSRLMLS	MENSRLMLS	MNSRLLAD	MENSRLLAD
with accession	Sn	0.6481	0.7789	0.7678	0.7992	0.7812
AF009962	Sp	0.6518	0.7890	0.7785	0.8075	0.7836
	Pr	0.5904	0.7806	0.7712	0.8004	0.7867

exon segments, convergence performance, computational complexity, and performance measures. The performance of various sign regressor–based maximum normalized AEPs including LMS-based AEPs in terms of specificity (Sp), sensitivity (Sn), and precision (Pr) is shown in Table 6.3.

More the value of all these metrics, the better the exon prediction accuracy for an adaptive algorithm. However, it also depends on the low computational complexity offered by an adaptive algorithm for better performance. From this table, AEP using MNSRLLAD has shown better performance for accurately locating exon segments compared to LMS-based AEP and all other sign regressor–based maximum normalized adaptive algorithms because of significantly low computational complexity compared to all other AEPs.

This algorithm also performs better in terms of performance measures than adaptive techniques used in Chapters 3–5. Therefore, the AEP using MNSRNLLAD could be useful for developing low complexity nanodevices of bioinformatics in applications based on system-on-a-chip (SOC) as well as lab-on-a-chip (LOC).

6.2 RECOMMENDATIONS FOR FUTURE RESEARCH

A large amount of data is being produced by a broad range of genetic techniques. Signal processing techniques, which already had a crucial impact on several areas, are a part of revolution in genetics because they can process large amounts of information rapidly and efficiently. To be more specific, accurate exon prediction in real DNA sequences continues to attract the attention of many scientists for different applications. The solution for many interesting bioinformatics issues is still unknown, and many issues still need better alternatives. In this book, we have developed and proved the effectiveness of the adaptive filtering techniques for improving the accuracy in exon prediction in real DNA sequences of *Homo sapiens* in bioinformatics applications. However, it is believed that the research published here can be extended considerably in different ways for possible future investigations. These are described shortly below:

- This book has concentrated on various AEPs for improving the accuracy in exon prediction in real DNA sequences. Further AEPs development is needed using bioinformatics- and cloud-based platforms, remote healthcare monitoring systems, etc.

- For practical implementation of the developed AEPs, they could be implemented on real-time embedded DSP processors like TI or field programmable gate arrays (FPGAs), etc.
- The low-complexity NSRLLAD-based AEPs from our research could be used to develop nanodevices of bioinformatics in applications related to SOC as well as LOC.
- New AEPs using neural networks and fuzzy logic can also be developed.
- Gene sequence analysis enables us to generate more efficient drugs and better food products and also to attain valuable insight into our own body functioning. Other areas of future interest may include extending our research on DNA sequences to produce disease-free crops, healthier farm animals, etc.
- Future scope of work related to bioinformatics can also be aimed at discovery of disease mechanisms, improved disease diagnosis, and prevention strategies to improve human health.

References

1. R. S. H. Istepanian, A. Sungoor, and J.-C. Nebel, "Comparative analysis of genomic signal processing for microarray data clustering", *IEEE Transactions on NanoBioscience*, vol. 10, no. 4, pp. 225–238, 2011.
2. K. Hoff, S. Lange, A. Lomsadze, M. Borodovsky and M. Stanke, "BRAKER1: Unsupervised RNA-seq-based genome annotation with genemark-ET and AUGUSTUS", *Bioinformatics*, vol. 32, no. 5, pp. 767–769, 2016.
3. V. V. Solovyev, A. A. Salamov, and C. B. Lawrence, "Identification of human gene structure using linear discriminant functions and dynamic programming", *Proceedings of 3rd International Conference on Intelligent Systems for Molecular Biology*, Cambridge, England, vol. 3, pp. 367–375, 1995.
4. I. M. Al-Turaiki, H. Mathkour, A. Touir, and S. Hammami, "Computational approaches for gene prediction: A comparative survey", *Proceedings of International Conference on Informatics Engineering and Information Science*, Kuala Lumpur, Malaysia, Part II, CCIS 252, Springer, pp. 14–25, 2011.
5. M. Borodovsky and A. Lomsadze, "Eukaryotic gene prediction using GeneMark.hmm-E and GeneMark-ES", *Current Protocols in Bioinformatics*, Chapter 4: Unit 4.6, pp. 1–10, 2011.
6. D. Kulp, D. Haussler, M. G. Reese, and F. H. Eeckman, "A generalized hidden Markov model for the recognition of human genes in DNA", *Proceedings of 4th International Conference on Intelligent Systems for Molecular Biology*, Halkidiki, Greece, vol. 4, pp. 134–142, 1996.
7. T. Inatsuki, K. Sato, and Y. Sakakibara, "Prediction of gene structures from RNA-seq data using dual decomposition", *IPSJ Transactions on Bioinformatics*, vol. 9, pp. 1–6, 2016.
8. M. Axelson-Fisk, *"Comparative Gene Finding: Models, Algorithms and Implementation"*, 2nd Edition, Springer, New York, 2015.
9. S. Harisha, *"Fundamentals of Bioinformatics"*, Chapter 9, 1st edition, I.K. International Publishing House Pvt. Ltd., New Delhi, pp. 200–205, 2010.
10. G. B. Singh, *"Fundamentals of Bioinformatics and Computational Biology"*, *Springer International Publishing*, Switzerland, 2015.
11. K. Murakami and T. Takagi, "Gene recognition by combination of several gene-finding programs", *Bioinformatics*, vol. 14, no. 8, pp. 665–675, 1998.
12. V. Pavlovic, A. Garg, and S. Kasif, "A Bayesian framework for combining gene predictions", *Bioinformatics*, vol. 18, no. 1, pp. 19–27, 2002.
13. G. N. Satapathi, P. Srihari, A. Jyothi, and S. Lavanya, "Prediction of cancer cell using DSP techniques", *Proceedings of IEEE International Conference on Communication and Signal Processing*, pp. 149–153, 2013.
14. E. Borrayo, E. G. Mendizabal-Ruiz, et.al. "Genomic Signal Processing Methods for Computation of Alignment-Free Distances from DNA Sequences", *PLOS One*, vol. 9, no. 1, pp. 1–13, 2014.
15. A. Travers and G. Muskhelishvili, "DNA structure and function", *The FEBS Journal*, vol. 282, no. 12, pp. 2279–2295, 2015.
16. J. Meidanis and C. Setubal, *"Introduction to Computational Molecular Biology"*, Cengage Learning, PWS Publishing Company, Boston, MA, 2007.
17. T. A. Brown, *"Gene Cloning and DNA Analysis: An Introduction"*, 7th edition, *Wiley-Blackwell Publishing*, Hoboken, NJ, 2016.

18. J. D. Watson and F. C. H. Crick, "Molecular structure of nucleic acids: A structure for deoxyribose nucleic acid", *Nature*, vol. 171, pp. 737–738, 1953.
19. R. J. Reece, *"Analysis of Genes and Genomes"*, 1st edition, *John Wiley & Sons Ltd.*, Hoboken, NJ; Cambridge University Press, Cambridge, 2004.
20. J. D. Watson, *"Molecular Biology of the Gene"*, 7th edition, *Pearson Education*, London, 2017.
21. J. W. Dale and M. von Schantz, *"From Genes to Genomes: Concepts and Applications of DNA Technology"*, 3rd edition, *Wiley-Blackwell Publishers*, Hoboken, NJ, 2011.
22. S. Bandyopadhyay, U. Maulik, and D. Roy, "Gene identification: Classical and computational intelligence approaches", *IEEE Transactions on Systems Man and Cybernetics*, vol. 38, no. 1, pp. 55–68, 2008.
23. NCBI BLAST, Basic Local Alignment Search Tool, (https://blast.ncbi.nlm.nih.gov/Blast.cgi), (accessed on October 20, 2014).
24. NCBI Reference Sequence Database (RefSeq), (https://www.ncbi.nlm.nih.gov/refseq/), (accessed on October 20, 2014).
25. EMBL-EBI Database, European Bioinformatics Institute Ensembl Genome Browser, (https://www.ebi.ac.uk/ena), (accessed on October 20, 2014).
26. M. Stanke, O. Schoffmann, B. Morgenstern, and S. Waack, "Gene prediction in eukaryotes with a generalized hidden Markov model that uses hints from external sources", *BMC Bioinformatics*, vol. 7, no. 62, pp. 1–11, 2006.
27. D. R. Kelley, B. Liu, A. L. Delcher, M. Pop, and S. L. Salzberg, "Gene prediction with Glimmer for metagenomic sequences augmented by classification and clustering", *Nucleic Acids Research*, vol. 40, no. 1, pp. 1–12, 2012.
28. E. Blanco, G. Parra, and R. Guigo, "Using geneid to identify genes", *Current Protocols in Bioinformatics*, vol. 18, pp. 4.3.1–4.3.28, 2007.
29. V. A. Makarov, "Computer programs for eukaryotic gene prediction", *Briefings in Bioinformatics*, vol. 3, no. 2, pp. 195–199. 2002.
30. A. Nagar, S. Purushothaman, and H. Tawfik, "Evaluation and fuzzy classification of gene finding programs on human genome sequences", *Proceedings of International Conference on Fuzzy Systems and Knowledge Discovery*, Atlanta, GA, pp. 821–829, 2005.
31. S. Logeswaran, E. Ambikairajah, and J. Epps, "A method for detecting short initial exons", *Proceedings of IEEE Workshop on Genomic Signal Processing and Statistics*, Texas, pp. 61–62, 2006.
32. Y. Saeys, P. Rouze, and Y. V. de Peer, "In search of the short ones improved prediction of short exons in vertebrates, plants, fungi and protists", *Bioinformatics*, vol. 23, no. 4, pp. 414–420, 2007.
33. W. Zhu, A. Lomsadze, and M. Borodovsky, "Ab initio Gene Identification in Metagenomic Sequences", *Nucleic Acids Research*, vol. 38, no. 12, pp. 1–15, 2010.
34. S. Foissac, P. Bardou, A. Moisan, M. J. Cros, and T. Schiex, "EuGene'Hom: A generic similarity-based gene finder using multiple homologous sequences", *Nucleic Acids Research*, vol. 31, no. 13, pp. 3742–3745, 2003.
35. A. Li, Z. Liu, K. Lezon-Geyda, S. Sarkar, D. Lannin, V. Schulz, I. Krop, E. Winer, L. Harris, and D. Tuck, "GPHMM: an integrated hidden Markov model for identification of copy number alteration and loss of heterozygosity in complex tumor samples using whole genome SNP arrays", *Nucleic Acids Research*, vol. 39, no. 12, pp. 4928–4941, 2011.
36. M. Alexandersson, S. Cawley, and L. Pachter, "SLAM cross-species gene finding and alignment with a generalized pair hidden Markov model", *Genome Research*, vol. 13, no. 3, pp. 496–502, 2003.
37. M. Ghorbani and H. Karimi, "Bioinformatics approaches for gene finding", *International Journal of Scientific Research in Science and Technology*, vol. 1, no. 2015, pp. 12–15, 2015.

38. G. Liu and Y. Luan, "Identification of protein coding regions in the eukaryotic DNA sequences based on marple algorithm and wavelet packets transform", *Abstract and Applied Analysis*, vol. 2014, pp. 1–14, 2015.
39. R. F. Voss, "Evolution of long-range fractal correlations and 1/f noise in DNA base sequences", *Physical Review Letters,* vol. 68, no. 25, pp. 3805–3808, 1992.
40. National Center for Biotechnology Information, (https://www.ncbi.nlm.nih.gov/), (accessed on January 25, 2014).
41. R. Zhang and C.-T. Zhang, "A brief review: The Z-curve theory and its application in genome analysis", *Current Genomics*, vol. 15, no. 2, pp. 78–94, 2014.
42. C. T. Zhang, R. Zhang, and H. Y. Ou, "The Z curve database: A graphic representation of genome sequences", *Bioinformatics*, vol. 19, no. 5, pp. 593–599, 2003.
43. A. Rushdi and J. Tuqan, "Gene identification using the Z-curve representation", *Proceedings of IEEE International Conference on Acoustics, Speech and Signal Processing (ICASSP'06)*, Toulouse, France, vol. 2, pp. 1024–1027, 2006.
44. P. D. Cristea, *"Genomic Signal Processing and Statistics"*, EURASIP Book Series on Signal Processing and Communications, Hindawi Publishing Corporation, London, vol. 2, pp. 15–65, 2005.
45. C. Cattani, "Complex representation of DNA sequences", Proceedings of International Conference on Bioinformatics Research and Development, Vienna, Austria, pp. 528–537, 2008.
46. W. R. Hamilton, "On quaternions", *Proceedings of the Royal Irish Academy*, vol. 3, pp. 1–16, 1847.
47. W. R. Hamilton, "*Elements of quaternions*", Nature, vol. 1556, no. 60, pp. 387–387, 1899.
48. A. K. Brodzik, "Quaternionic periodicity transform an algebraic solution to the tandem repeat detection problem", *Bioinformatics*, vol. 23, pp. 694–700, 2007.
49. T. M. Inbamalar and R. Sivakumar, "Improved algorithm for analysis of DNA sequences using multiresolution transformation", *The Scientific World Journal*, vol. 2015, Article ID 786497, pp. 1–9, 2015.
50. S. S. Nair and T. Mahalakshmi, "Visualization of genomic data using inter-nucleotide distance signals", *Proceedings of IEEE International Conference on Genomic Signal Processing*, Philadelphia, PA, pp. 1–4, 2005.
51. R. Arora and W. A. Sethares, "Detection of Periodicities in gene sequences a maximum likelihood approach", *Proceedings of IEEE International Workshop on Genomic Signal Processing and Statistics*, Houston, TX, pp. 1–4, 2007.
52. L. W. Ning, H. Lin, H. Ding, J. Huang, N. Rao and F. B. Guo, "Predicting bacterial essential genes using on sequence composition information", *Genetics and Molecular Research*, vol. 13, no, 2014, pp. 4564–4572, 2014.
53. S. Tiwari, S. Ramachandran, A. Bhattacharya, S. Bhattacharya and R. Ramaswamy, "Prediction of probable genes by Fourier analysis of genomic sequences", *Computer Applications in the Biosciences (CABIOS)*, vol. 13, no. 3, pp. 263–270, 1997.
54. S. L. Salzberg, M. Pertea, A. L. Delcher, M. J. Gardner, and H. Tettelin, "Interpolated Markov models for eukaryotic gene finding", *Genomics*, vol. 59, no. 1, pp. 24–31, 1999.
55. B. Morgenstern, O. Rinner, S. Abdeddaïm, D. Haase, K. F. Mayer, A. W. Dress, and H. W. Mewes, "Exon discovery by genomic sequence alignment", *Bioinformatics*, vol. 18, no. 6, pp. 777–787, 2002.
56. P. P. Vaidyanathan and B.-J. Yoon, "Digital filters for gene prediction applications", *Proceedings of IEEE International Conference on Circuits, Systems and Computers*, Oranjestad, Aruba, Netherlands, pp. 1–5, 2002.
57. S.S. Sahu, and G. Panda, "An efficient signal processing approach in eukaryotic gene prediction", *Proceedings of Eighth Asia Pacific Bioinformatics Conference Asia Pacific Bioinformatics Conference, BMC Bioinformatics*, Bangalore, India, pp. 1–12, 2010.

58. M. S. Mabrouk, S. M. Naeem, and M. A. Eldosoky, "Different genomic signal processing methods for eukaryotic gene prediction: A systematic review", *Biomedical Engineering: Applications, Basis and Communications*, vol. 29, no. 01, pp. 1730001-1–1730001-18, 2017.
59. D. DeCaprio, J. P. Vinson, M. D. Pearson, P. Montgomery, M. Doherty, and J. E. Galagan, "Conrad: Gene prediction using conditional random fields", *Genome Research*, vol. 17, no. 9, pp. 1389–1398, 2007.
60. H. Saberkari, M. Shamsi, H. Heravi, M. H. Sedaaghi, "A novel fast algorithm for exon prediction in eukaryotes genes using linear predictive coding model and goertzel algorithm based on the Z-Curve", *International Journal of Computer Applications*, vol. 67, no. 2013, pp. 25–38, 2013.
61. T. M. Inbamalar and R. Sivakumar, "Study of DNA sequence analysis using DSP techniques", *Journal of Automation and Control Engineering*, vol. 1, pp. 336–342, 2013.
62. E. N. Trifonov, "3-, 10.5-, 200- and 400-base periodicities in genome sequences", *Physica A*, vol. 249, pp. 511–516, 1998.
63. H. Herzel, O. Weiss, and E. N. Trifonov, "10–11 bp periodicities in complete genomes reflect protein structure and DNA folding", *Bioinformatics*, vol. 15, no. 3, pp. 187–193, 1999.
64. J. W. Fickett and C. S. Tung, "Assessment of protein coding measures", *Nucleic Acids Research*, vol. 20, no. 24, pp. 6441–6450, 1992.
65. M. N. Mohanty, "A signal processing approach for eucaryotic gene identification", *ICT and Critical Infrastructure: Proceedings of the 48th Annual Convention of Computer Society of India- Volume II*, pp. 361–370, 2014.
66. S. T. Eskesen, F. N. Eskesen, B. Kinghorn, and A. Ruvinsky, "Periodicity of DNA in exons", *BMC Molecular Biology*, vol. 5, no. 12, pp. 1–11, 2004.
67. M. Akhtar, "Genomic Sequence Processing: Gene Finding in Eukaryotes", *Ph.D Thesis, The University of New South Wales*, Sydney NSW, Australia, 2008. (last viewed on 12 July, 2015)
68. S. Datta and A. Asif, "A fast DFT-based gene prediction algorithm for identification of protein coding regions", *Proceedings of IEEE International Conference on Acoustics, Speech, and Signal Processing*, vol. 5, pp. 653–656, 2005.
69. S. Verma, "Detection of protein coding regions using goertzel algorithm", *International Journal of Computer Applications*, vol. 124, no. 2, pp. 1–4, 2015.
70. M. Akhtar, J. Epps and E. Ambikairajah, "Paired spectral content measure for gene and exon prediction in eukaryotes", *Proceedings of IEEE International Conference on Information and Emerging Technologies*, Karachi, Pakistan, pp. 1–4, 2007.
71. D. Kotlar and Y. Lavner, "Gene prediction by spectral rotation measure: A new method for identifying protein-coding regions", *Genome Research*, vol. 18, pp. 1930–1937, 2003.
72. M. Ahmad, A. Abdullah and K. Buragga, "A novel optimized approach for gene identification in DNA sequences", *Journal of Applied Sciences*, vol. 11, pp. 806–814, 2011.
73. T. T. Tran, V. A. Emanuele II, and G. Tong Zhou, "Techniques for detecting approximate tandem repeats in DNA", *Proceedings of IEEE International Conference on Acoustics, Speech, and Signal Processing*, Montreal, Canada, pp. 449–452, 2004.
74. H. M. Wassfy, M. M. Abd Elnaby, M. L. Salem, M. S. Mabrouk, and A.-A. A. Zidan, "Advanced DNA mapping schemes for exon prediction using digital filters", *American Journal of Biomedical Engineering*, vol. 6, no. 1, pp. 25–31, 2016.
75. N. Rao and S. J. Shepherd, "Detection of 3-periodicity for small genomic sequences based on AR techniques", *Proceedings of International Conference on Communication, Circuits and Systems*, vol. 2, pp. 1032–1036, 2004.

76. N. Y. Song and H. Yan, "Autoregressive modeling of DNA features for short exon recognition", *2010 IEEE International Conference on Bioinformatics and Biomedicine (BIBM)*, Hong Kong, pp. 450–455, 2010.
77. B. Ma, D. Qu, and Y.-S. Zhu, "A novel adaptive filtering approach for genomic signal processing", *Proceedings of IEEE 10th International Conference on Signal Processing Proceedings*, Beijing, China, pp. 1–5, 2010.
78. S. O. Haykin, *Adaptive Filter Theory*, 5th edition, *Pearson Education Ltd.*, London, 2014.
79. E. R. Dougherty, I. Shmulevich, and M. L. Bittner, "Genomic signal processing the salient issues", *EURASIP Journal on Applied Signal Processing*, vol. 2004, no. 1, pp. 146–153, 2004.
80. H. Saberkari, M. Shamsi, H. Heravi, and M. H. Sedaaghi, "A fast algorithm for exonic regions prediction in DNA sequences", *Journal of Medical Signals and Sensors*, vol. 3, no. 3, pp. 139–149, 2013.
81. E. Walach and B. Widrow, "The least mean fourth (LMF) adaptive algorithm and its family", *IEEE Transactions on Information Theory*, vol. 30, no. 2, pp. 2275–283, 1984.
82. H. ChaeWoo, "Variable step size LMS algorithm using squared error and autocorrelation of error", *Procedia Engineering*, vol. 41, pp. 47–52, 2012.
83. Y.-P. Li, T.-S. Lee, and B.-F. Wu, "A variable step-size sign algorithm for channel estimation", *Signal Processing*, vol. 102, pp. 304–312, 2014. Elsevier.
84. E. Eweda, "Global stabilization of the least mean fourth algorithm", *IEEE Transactions on Signal Processing*, vol. 60, no. 3, pp. 1473–1477, 2012.
85. F. Chen, T. Shi, S. Duan, L. Wang, and J. Wu, "Diffusion least logarithmic absolute difference algorithm for distributed estimation", *Signal Processing*, vol. 142, pp. 423–430, 2018.
86. M. Li, Q. Li, G. U. Ganegoda, J. X. Wang, F. X. Wu, and Y. Pan, "Prioritization of orphan disease-causing genes using topological feature and go similarity between proteins in interaction networks", *Science China Life Sciences*, Vol. 57, pp. 1064–1071, 2014.
87. J. E. Dickerson, A. Zhu, D. L. Robertson, and K. E. Hentges, "Defining the role of essential genes in human disease", *PloS One*, vol. 6, pp. 1–10, 2011.
88. S. Cole, "Comparative myco bacterial genomics as a tool for drug target and antigen discovery", *The European Respiratory Journal*, vol. 20, no. 36, pp. 78s–86s, 2002.
89. S. Maji and D. Garg, "Progress in gene prediction: Principles and challenges", *Current Bioinformatics*, vol. 8, pp. 226–243, 2013.
90. O. Abbasi, A. Rostami, and G. Karimian, "Identification of exonic regions in DNA sequences using cross-correlation and noise suppression by discrete wavelet transform", *BMC Bioinformatics*, vol. 12, pp. 1–10, 2011.
91. N. Goel, S. Singh, and T. C. Aseri, "A review of soft computing techniques for gene prediction", *ISRN Genomics*, vol. 2013, pp. 1–8, 2013.
92. S. Putluri, and S. Y. Fathima, "Novel adaptive exon predictor for DNA analysis using singular value decomposition", *International Journal of Measurement Technologies and Instrumentation Engineering (IJMTIE)*, vol. 6, no. 1, pp. 1–11, 2017.
93. A. D. Poularikas, *"Adaptive Filtering: Fundamentals of Least Mean Squares with MATLAB"*, *CRC Press, Taylor and Francis Group*, Boca Raton, FL, 2015.
94. R. H. Kwong and E. W. Johnston, "A variable step size LMS algorithm", *IEEE Transactions on Signal Processing*, vol. 40, no. 7, pp. 1633–1642, 1992.
95. Y. Azuma and S. Onami, "Automatic cell identification in the unique system of invariant embryogenesis in caenorhabditis elegans", *Biomedical Engineering Letters*, vol. 4, no. 2014, pp. 328–337, 2014.

96. V. J. Mathews and S.-H. Cho, "Improved convergence analysis of stochastic gradient adaptive filters using the sign algorithm", *IEEE Transactions on Acoustic, Speech, Signal Processing*, vol. 35, no. 4, pp. 450–454, 1987.
97. B. Mulgrew and C. F. Cowan, *"Adaptive Filters and Equalisers", Kluwer Academic Publishers, Springer*, New York, 2012.
98. E. Eweda, "A new performance measure of the normalized lms adaptive filter", *Proceedings of IEEE International Conference on Signal Processing and Communications*, Dubai, United Arab Emirates, pp. 652–658, 2007.
99. A. H. Sayed, *"Fundamentals of Adaptive Filtering"*, John Wiley and Sons, *New York*, 2010.
100. S. Zhang and W. X. Zheng, "Normalized least mean-square algorithm with variable step size based on diffusion strategy", *Proceedings of IEEE International Symposium on Circuits and Systems (ISCAS)*, Phagwara, India, pp. 1–4, 2018.
101. A. D. Poularikas, *"Understanding Digital Signal Processing with MATLAB and Solutions", CRC Press*, Boca Raton, FL, 2017.
102. M. Rupp and R. Frenzel, "Analysis of LMS and NLMS Algorithms with delayed coefficient update under the presence of spherically invariant process", *IEEE Transactions on Signal Processing*, vol. 42, no. 3, pp. 668–672, 1992.
103. L. R. Vega and H. Rey, "A rapid introduction to adaptive filtering", *Springer*, New York, 2013.
104. S. C. Douglas and T. H. Y. Meng, "Normalized data nonlinearities for LMS adaptation", *IEEE Transactions on Signal Processing*, vol. 42, pp. 1352–1365, 1994.
105. S. Jo and S. W. Kim, "Consistent normalized least mean square filtering with noisy data matrix", *IEEE Transactions on Signal Processing*, vol. 53, no. 6, pp. 2112–2123, 2005.
106. A. Uncini, *"Fundamentals of Adaptive Signal Processing", Springer International Publishing*, Switzerland, 2015.
107. W.-P. Ag and B. Farhang-Boroujeny, "A new class of gradient adaptive step-size LMS algorithms", *IEEE Transactions on Signal Processing*, vol. 49, no. 4, pp. 805–810, 2001.
108. Y. Zhang, N. Li, J. A. Chambers and Y. Hao, "New Gradient-Based Variable Step Size LMS Algorithms", *Eurasip Journal on Advances in Signal Processing*, Article ID 529480, vol. 2008, pp. 1–9, 2008.
109. J. M. Grizz, J. Ramrej, S. C. Alvarez, C. G. Puntonet, E. W. Lang, and E. Deniz, "A novel LMS algorithm applied to adaptive noise cancellation", *IEEE Signal Processing Letters*, Vol. 16, pp. 34–37, 2009.
110. S. Dhull, S. Arya and O. P. Sahu, "Performance variation of LMS and its different Variants", *International Journal of Computer Science and Security, (IJCSS)*, vol. 4, no. 5, pp. 491–496, 2010.
111. B. Farhang-Boroujeny, *"Adaptive Filters-Theory and Applications"*, 2nd Edition, *John Wiley and Sons*, Chichester, 2013.
112. P. S. R. Diniz, *"Adaptive Filtering, Algorithms and Practical Implementation"*, 4th Edition, *Springer*, New York, 2013.
113. E. Eweda, "Comparison of RLS, LMS, and sign algorithms for tracking randomly time-varying channels", *IEEE Transactions on Signal Processing*, vol. 42, no. 11, pp. 2937–2944, 1994.
114. S. C. Douglas, "A family of normalized LMS algorithms", *IEEE Signal Processing Letters*, vol. 1, pp. 1352–1365, 1994.
115. W. Kenneth Jenkins, A. W. Hull, J. C. Strait, B. A. Schnaufer, and X. Li, *"Advanced Concepts in Adaptive Signal Processing", Springer Science Publishers*, New York, 2012.
116. A. Zerguine, "Convergence and steady-state analysis of the normalized least mean fourth algorithm", *Digital Signal Processing*, vol. 17, no. 1, pp. 17–31, 2007.

117. P. I. Hubscher, J. C. M. Bermudez and V. H. Nascimento, "A mean-square stability analysis of the least mean fourth adaptive algorithm", *IEEE Transactions on Signal Processing*, vol. 55, no. 8, pp. 4018–4028, 2007.
118. E. Eweda and N. Bershad, "Stochastic analysis of a stable normalized least mean fourth algorithm for adaptive noise canceling with a white Gaussian reference", *IEEE Transactions on Signal Processing*, vol. 60, no. 12, pp. 6235–6244, 2012.
119. S. M. Jung, J. H. Seo, and P. G. Park. "A variable step-size diffusion normalized least-mean-square algorithm with a combination method based on mean square deviation", *Circuits, Systems and Signal Processing*, vol. 34, no. 10, pp. 3291–3304, 2015.
120. L. G. Morales, *"Adaptive Filtering"*, InTech Publishers, 2011.
121. N. Li, Y. Zhang, Y. Hao and J. A. Chambers, "A new variable step-size NLMS algorithm designed for applications with exponential decay impulse responses", *Signal Processing*, vol. 88, pp. 2346–2349, 2008.
122. A. I. Sulyman and A. Zerguine, "Convergence and steady-state analysis of a variable step-size NLMS algorithm", *Signal Processing*, vol. 83, pp. 1255–1273, 2003.
123. S. Zhang, J. Zhang, and H. Han, "Robust shrinkage normalized sign algorithm in impulsive noise environment", *IEEE Transactions on Circuits and Systems II: Express Briefs*, vol. 64, no. 1, pp. 91–95, 2017.
124. M. O. Sayin, N. Denizcan Vanli, S. S. Kozat, "A novel family of adaptive filtering algorithms based on the logarithmic cost", *IEEE Transactions on Signal Processing*, vol. 62, no. 17, pp. 4411–4424, 2014.
125. M. K. Ahirwal, A. Kumar, et al., "Performance prediction of adaptive filters for EEG signal", *IET Science Measurement and Technology*, vol. 11, no. 5, pp. 525–531, 2017.
126. S. Mula, V. C. Gogineni, and A. S. Dhar, "Algorithm and architecture design of adaptive filters with error nonlinearities", *IEEE Transactions on Very Large Scale Integration Systems*, vol. 25, no. 9, pp. 2588–2601, 2017.
127. V. C. Gogineni and S. Mula, "A family of constrained adaptive filtering algorithms based on logarithmic cost", *IEEE Transactions on Signal Processing*, vol. 62, pp. 1–14, 2017.
128. M. O. Sayin, N. Denizcan Vanli, S. S. Kozat, "Improved convergence performance of adaptive algorithms through logarithmic cost", *Proceedings of IEEE International Conference on Acoustic, Speech and Signal Processing (ICASSP)*, Florence, Italy, pp. 4513–4517, 2014.

Index

Note: **Bold** page numbers refer to tables; *italic* page numbers refer to figures